WATER CHALLENGES OF AN URBANIZING WORLD

Edited by **Matjaž Glavan**

Water Challenges of an Urbanizing World
http://dx.doi.org/10.5772/intechopen.68339
Edited by Matjaž Glavan

Contributors

Megersa Olumana Dinka, Suat Irmak, Denis Mutiibwa, Ayse Kilic, Alejandra A. López-Caloca, Felipe Omar Tapia-Silva, Guadalupe Rivera, Amira Abdelrasoul, Heloisa Westphalen, Adejumoke Abosede Inyinbor, Babatunde Olufemi Adebesin, Tabitha Adelani-Akande, Adewumi Dada, Toyin Oreofe, Abimbola Oluyori, Xiang Gao, Tengfei Yu, Shawn Larson, Amy Y. Green, Angela Smith, Susana Neto, Anita Rakić, Shailendra K. Saxena

Published by InTech
Janeza Trdine 9, 51000 Rijeka, Croatia

Publishing Process Manager Edi Lipovic
Technical Editor SPi Global
Cover InTech Design team

First published March, 2018
Printed in Croatia
Legal deposit, Croatia: National and University Library in Zagreb

Additional hard copies can be obtained from orders@intechopen.com

Water Challenges of an Urbanizing World, Edited by Matjaž Glavan
p. cm.
Print ISBN 978-953-51-3893-8
Online ISBN 978-953-51-3894-5

For EU product safety concerns:
IN TECH d.o.o., Prolaz Marije Krucifikse Kozulić 3, 51000 Rijeka, Croatia,
info@intechopen.com or visit our website at intechopen.com.

Meet the editor

Dr. Matjaž Glavan is an assistant professor of Environmental Planning at Biotechnical Faculty of the University of Ljubljana and the chair for Agrometeorology, Agricultural Land Management, Economics and Rural Development. He completed his BSc degree in Agronomy at the University of Ljubljana, Slovenia, and his MSc degree in Integrated Catchment Management at Cranfield University, UK. He obtained his PhD degree in Biological and Biotechnological Sciences at the University of Ljubljana in 2011. His main field of expertise is integrated water and land management in relation to agriculture. His research focus is also on modelling spatial, economic and environmental impacts of land-use changes and agri-environmental measures on water, soil and land resource quality and quantity. He is also specialized in catchment modelling with the Soil and Water Assessment Tool.

Contents

Preface

Global water crisis is a challenge to the security, political stability and environmental sustainability of developing nations and with climate, economically and politically, induces migrations also for the developed ones. Currently, the urban population is 54% with prospects that by the end of 2050 and 2100 66% and 80%, respectively, of the world's population will live in urban environment. This opens questions about provision of proper quantity and quality of fresh water and food. Water presents a challenge as it is the foundation of life. Untreated water abstracted from polluted resources and destructed ecosystems as well as discharge of untreated waste water is the cause of health problems and death for millions around the globe. In all countries around the globe, governments, engineers and scientists are faced with the challenge of how to supply proper quantities of safe drinking water. Competition for water is wide among agriculture, industry, power companies and recreational tourism as well as nature habitats. Climate changes with changed spatial and temporal distribution of precipitation are a major threat to the water resources. This book intends to provide the reader with a comprehensive overview of the current state of the art in integrated assessment of water resource management in the urbanizing world, which is a foundation to develop society with secure water availability, food market stability and ecosystem preservation.

This book entitled *Water Challenges of an Urbanizing World* covers issues related to water resource quantity and quality, safe drinking water, drinking water supply systems, water pollution, water monitoring, water governance, water legislation, decision support system, population growth, barriers for drinking water availability, climate change and water infrastructure in relation to urban development. The book is structured in two sections, which address different aspects of those issues, consisting of ten chapters.

The first section, "Water Resource Quality and Quantity", presents five chapters. Chapter 1 focuses on the neglected tropical waterborne infectious diseases including strategies for mitigation, prevention and integrated control including vaccine and drugs. Chapter 2 addresses microbial hazards (i.e., *Legionella* bacterium), which make the largest contribution to waterborne disease in developed and developing countries. The control of the disease has been one of the major problems in countries with low- and middle-income economies, including Croatia. Chapter 3 reviews water pollution effects vis-a-vis global challenges, threat and climatic impacts and various possible preventive measures. Chapter 4 is a case study of the site located on Elliot Bay in Puget Sound in Seattle, WA, USA, that presents a descriptive temporal analysis of variability in the incoming saltwater to Seattle Aquarium. Chapter 5 reviews key sources of microplastics, assessing their capacity to problematize water resources in urban environments, offers analysis of its effects on the environment and suggests methodologies to control it.

The second section, "Water Management", presents five chapters. Chapter 6 addresses the need for an integrated water management and governance in the urban areas that will more effectively understand the multiple dimensions and complexity of this approach in coherent relation with land-use planning. Chapter 7 is a case study of Gansu section of Western Longhai-Lanxin Economic Zone, China, which presents impacts of spatiotemporal variations of water resources as a limiting factor for urban development. Chapter 8 is a case study of Grijalva-Usumacinta basin in the state of Tabasco, Mexico, which shows advances in the field of water body monitoring procedures with radar images to define the extension and frequency of inundation for continental waters. Chapter 9 is a case study of High Plains in the USA that presents analysis for detecting summer air temperature change by investigating trends of two separate climate periods of 1895–1930 and 1871–2006. Chapter 10 presents aspects of safe drinking water and definitions of water safety, water access, benefits, basic principles and regulations of safe drinking water supplies, potential factors challenging the sustainable supply and water quality standards and parameters.

Matjaž Glavan, PhD
Assistant Professor in Environmental Planning
Biotechnical Faculty of the University of Ljubljana, Slovenia

Section 1

Water Resources Quality and Quantity

Chapter 1

Introductory Chapter: Neglected Tropical Waterborne Infectious Diseases - Strategies for Mitigation

Shailendra K. Saxena, Swatantra Kumar,
Amrita Haikerwal and Madan L.B. Bhatt

Additional information is available at the end of the chapter

http://dx.doi.org/10.5772/intechopen.74322

1. Introduction

Uninterrupted rise in population growth and human development has exerted diverse pressure on the quality and quantity of water with privilege of access to them. The interface of the human health and water experiences the most immense pressure due to the emergence of water-associated pathogens. Water-associated infectious diseases are the principal cause of morbidity and mortality in the developing and underdeveloped countries. Interplay between pathogen and its environment is complex and crucial for understanding the survival of the species under selective conditions. Potential drivers for the emergence and reemergence of water-associated infectious diseases can be explained by environmental aberration, implementation of newer technologies, scientific advances and changes in human behavior and vulnerability [1]. Understanding the plausible cause of the emergence of water-associated pathogens has become the national and international priority areas, which is fundamental for drinking water treatment and water safety management [2]. Sanitation and hygiene are vital for human health, development and survival. Poor sanitation is coupled with huge range of diseases and approximately 2.4 billion people are underprivileged for advanced sanitation [3]. Many countries are able to provide adequate sanitation to their population, leaving people at risk for water, sanitation and hygiene (WASH)-related diseases. Fundamental source of emergence of water-associated infection is coupled with poor sanitation where contaminated fecal water resources are most predominant [4]. Waterborne infections are caused by either exposure to or ingestion of contaminated water and hence strongly coupled with scarcity, availability and access to water. Environmental factors, such as heavy rainfall, flooding, drought and temperature variation, may contribute for the emergence of water-associated pathogens [5]. Heavy rainfall and flooding are responsible for huge pathogen load in naturally existing water resources. Certainly, heavy rainfall or flooding causes the overwhelming flow of sewage treatment plants, dissipated animal-generated manures and redistribution or remobilization of contaminated water sediments [6].

Various kind of microbial pathogens are emerging from water and can be classified into specific categories. The most significant ones are arises from fecal contaminated water sources with high number of viruses, bacteria and parasites [7]. Based on the origin of the etiological agents, water-associated infectious diseases have been well categorized into five disease groups. So far, the water-associated diseases have been the major priority area of national and international government agencies for access of quality water for drinking and other uses. However, some of the water-associated pathogens fall into the category of neglected tropical diseases (NTDs), which need further attention with improved management system [8]. This chapter focuses on the neglected tropical waterborne infectious diseases including strategies for mitigation, prevention and integrated control such as vaccine and drugs.

2. Classification of water-associated pathogens

Water-associated infectious diseases can be broadly classified as water-borne, water-based, water-dispersed, water-related and water-washed [9]. Water-borne diseases can be described as microorganisms emerged from consumption of fecal-contaminated water and can be divided into four subcategories, including enteroviruses, bacteria, protozoa and helminthes [10]. Enormous range of diseases are caused by water-borne infection where foremost diseases are typhoid fever, cholera, bacillary dysentery, gastroenteritis, leptospirosis, poliomyelitis, aseptic meningitis, infectious hepatitis, amoebic meningoencephalitis and ascariasis [11]. Unlike water-borne, water-based etiological agents are primarily worms that spend a part of their life cycle in water bodies. The best known classical example of water-based infectious disease is schistosomiasis [12] caused by a human contact with water infested with the larva of some parasitic worms known as cercariae. Water-dispersed incorporates infectious etiological agents which proliferate in water bodies and may enter into the human body through respiratory tract. The most prominent water-dispersed disease is Legionellosis (Pontiac fever) caused by *Legionella* results in pneumonia [13]. Water-related infectious diseases include vector-borne diseases where the vector inhabitants and the related intensity of the disease or outbreak depend on the presence of environmental water [14]. Burden of infection dynamics of water-related infectious diseases is linked with hydroclimatological drivers across a broad range of spatial and temporal scale. Water-washed diseases can be caused by poor personal and domestic hygiene due to clean water deprivation resulting in contact of eye or skin with contaminated water. A broad range of diseases can be assigned in this category including shigellosis, trachoma, leprosy, skin infections, ulcers, scabies and conjunctivitis [15]. Despite the well-categorized water-associated diseases, few of them have been grouped as neglected tropical waterborne infectious diseases, including trachoma, schistosomiasis, leptospirosis, human African trypanosomiasis, dengue, lymphatic filariasis and many others, which have been listed in **Table 1**.

Category	Diseases	Pathogens	Vaccine	Drugs
Water-washed	Trachoma	*Chlamydia trachomatis*	—	Trimethoprim-sulfonamide combinations, amoxicillin, macrolides, chloramphenicol and tetracycline
Water-based	Schistosomiasis	*Schistosoma*	Sh-28-GST, Sm-14, schistosome protein- calpain (Sm-p80)	Praziquantel
Waterborne	Leptospirosis	*Leptospira*	—	Cephalosporins, cefotaxime and ceftriaxone and penicillin G
Water-related	Lymphatic filariasis	*Wuchereria bancrofti*	—	Mectizan, diethylcarbamazine and albendazole
Water-based	Onchocerciasis	*Onchocerca volvulus*	—	Ivermectin
Water-related	Dengue	Dengue virus	Dengvaxia, DENVax, LATV $\Delta 30$	Suramin, lovastatin, pentosan polysulfate, castanospermine (CSP) and deoxynojirimycin
Water-related	African trypanosomiasis	*Trypanosoma brucei*		Pentamidine, suramin, melarsoprol, eflornithine, nifurtimox
Water-based	Dracunculiasis	Guinea worm	—	—

Table 1. Neglected tropical water-borne infectious diseases, category, available vaccine and drugs.

3. Epidemiology

A global database for water-associated pathogens and diseases has been established based on Global Infectious Diseases and Epidemiology Network (GIDEON). The infection dynamics and burden of water-associated infectious diseases can be linked to hydroclimatological drivers across a wide range of spatial and temporal regions [16]. Disease prevalence can be significantly correlated with the exposure and transmission rate, which are affected by environmental factors such as rainfall, air/water temperature and seasonal variability. For example, flood in an area may cause the contamination of drinking water sources with raw sewage and washing out of the etiological agents from open defecation sites [17]. In general, rainfall may affect human activities related to personal hygiene and water contacts potentially increasing the transmission risk. Drinking water and recreational settings are the most prominent source of water-associated infections.

The epidemiology of water-associated NTDs has been discussed here. A water-washed bacterium *Chlamydia trachomatis* causes trachoma. The bacterium is transmitted through personal contact and causes trichiasis if left untreated. The disease is mostly prevalent in areas with poorly developed water sanitation facility with almost 8 million people visually impaired and approximately 500 million people at the risk of infection [18]. Water-based NTD, schistosomiasis, caused by a human contact with water infested with the larva of parasitic worms known as *cercariae.* Approximately, 206.5 million people required treatment for schistosomiasis in year 2016 and around 700 million people are at risk [19]. So far, three schistosome species have been documented including *Schistosoma haematobium* known to cause urogenital infection and reported primarily in sub-Saharan Africa, *Schistosoma japonicum* known for intestinal infection and has been reported in Indonesia, Philippines and People's republic of China and *Schistosoma mansoni* which also infects intestine and has been reported in Brazil, Caribbean island and sub-Saharan Africa. A water-related, human African trypanosomiasis (HAT), commonly known as sleeping sickness, is caused by parasitic protozoa known as trypanosomes. The disease transmission is prevalent in sub-Saharan African countries where it is known to transmit through tsetse fly. From last 50 years, significant reduction in the number of new cases has been achieved in year 2009 where the number of reported cases was less than 10,000 [20]. Another imperative water-related NTD is dengue with an estimated 500,000 people with severe dengue necessity of hospitalization annually and approximately 2.5% of them die. According to WHO, around 128 countries are at risk for dengue infection with an estimated 3.9 billion people [21]. Leptospirosis is endemic in several countries of South-East Asia where most of the human infection reported from India, Sri Lanka, Indonesia and Thailand during the rainy season. So far, major outbreaks have been reported in Jakarta (2003), Mumbai (2005) and Sri Lanka (2008). Current available reports are suggesting that the global incidence rate is ranging from 0.1 to 10 in 100,000 annually [22]. Lymphatic filariasis caused by a parasite *Wuchereria bancrofti,* which is responsible for 90% of the cases. WHO launched Global Programme to Eliminate Lymphatic Filariasis (GPELF) in year 2000, and currently, 856 million people in 52 countries are living in areas which need preventive chemotherapy to reduce the transmission level [23].

4. Strategies for mitigation

The estimated 1 billion of the world's poorest population is affected by NTDs that comprise parasitic, bacterial and viral agents causing 534,000 deaths per year [24]. The NTDs are highly endemic in rural regions and in low-income countries with an average income less than US$2 a day. This weaker section is unable to make any attempt in improving their children growth and development, education and malnourishment. As the target of NTDs is the poverty-stricken people, the market for drugs and vaccines are not well supported by the pharmaceutical industry, as well as the number of drugs available for treatment of NTDs is also limited and unchanged since the middle of the twentieth century. The preventive approaches suggested by WHO for controlling NTDs are case management, preventive chemotherapy, vector control, improved quality water and sanitation and zoonotic disease management [25].

The leading health organizations, such as WHO, CDC and UNICEF, have initiated various sustainable strategies across the globe to provide 2.3 billion unprivileged people access to basic sanitation, hygiene and clean water (**Figure 1**). International Water Association (IWA) and WHO has introduced Water Safety Plans (WSPs) that ensures the safety of water from its procurement to delivering to the consumers in urban and rural regions of 72 countries [26].WASH is a global program strengthened with a vision to ensure safe drinking water and

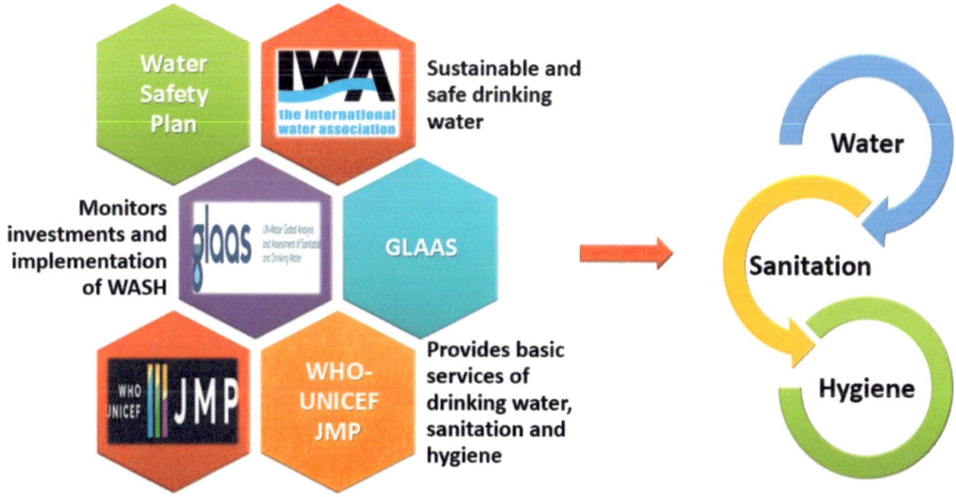

Figure 1. Global schemes and strategies for implementation of WASH. The enforcement of global schemes and strategies such as Water Safety Plan, GLAAS and WHO-UNICEF JMP will provide sustainable management of WASH.

maintenance of sanitation and hygiene to unprivileged people. The implementation of WASH and prevention of NTDs are integrated [27]. Global Analysis and Assessment of Sanitation and Drinking-Water (GLAAS) was initiated by UN and WHO to monitor the implementation of WASH [28].

The systematic maintenance of WASH may prevent and control water-borne NTDs such as trachoma, dengue fever, dracunculiasis, lymphatic filariasis, schistosomiasis, soil-transmitted helminthes and onchocerciasis (river blindness). WHO commenced the Global Elimination of Trachoma by 2020 (GET 2020) with a vision to launch the SAFE strategy that links four components such as surgery, antibiotics, facial cleanliness and environmental improvements (SAFE) to tackle trachoma [29]. This infectious disease is accountable for 3% of world's blindness, where an estimated 8 million people are irreversibly visually impaired and another 84 million cases are in need of treatment [30]. Likewise, WHO and UNICEF jointly monitor a program called the comprehensive strategy for the control and prevention of worms such as soil-transmitted helminthes (STH)—intestinal worms such as ascariasis, hookworm and trichuriasis [31]. Children without worms (CWW) advocate about maintenance of water, sanitation, hygiene education, deworming (WASHED) in order to discontinue the cycle of STH reinfection [32]. The World Health Assembly (WHA) in May 2012 has urged to eliminate schistosomiasis by providing essential drugs, practicing safe WASH and focusing on health education [33]. Dracunculiasis (Guinea worm), another waterborne NTD, has been adopted by WHA with a goal to achieve complete absence of Guinea worm disease cases for 3 or more years. Later, Guinea Worm Eradication Program was lead in partnership with Centers for Disease Control and Prevention (CDC), UNICEF and the Carter Centre [34]. Lymphatic filariasis and onchocerciasis are also targeted in WHO 2020 for eradication of NTDs. The London declaration to combat NTDs has estimated that US$785 million has been donated to NTD programs to strengthen and support drug availability and its distribution and research and development [35].

5. Prevention and integrated control

With the rising number of cases of water-associated infectious disease, there is an urgent need to regularly monitor the basic services of safe drinking water, sanitation and hygiene. The water resources are rivers, lakes and streams that replenish the groundwater table, and further, the water is stored in high-capacity storage tanks or else are directly delivered at the dwellings via piped or nonpiped facilities. The contamination of water resources due to open defecation near rivers, washing clothes, infiltration of solid wastes and chemicals from industries or bathing of animals like buffaloes deteriorates the quality of water. The contamination-free drinking water is the basic necessity for one and all without any inequality. The processing and pretreatment of the unsafe adulterated water from the source or because of the handling and storage issues could be used for drinking or in food preparation [36]. Numerous methods in combinations are used to minimize the microbial load of drinking water such as aeration, storage and sedimentation of water, filtration, disinfection of water and UV rays exposure. Various awareness programs regarding the basic hygiene facilities are necessary to be promoted to educate the population living especially in rural areas of developing countries. The global burden of mosquito-transmitted infectious diseases is rising at an alarming pace. The contaminated stagnant water kept in utensils for storage purpose is the dwelling of mosquito eggs and larvae. The mosquito-vector control program lists various methods such as avoiding water logging, repellents, insecticide-treated impregnated nets, larvivorous fish *Gambusia affinis*, and disinfecting their breeding habitats [37]. The use of toilets in every household and public place will end open defecation and prevent fecal contamination of water [38].

6. Vaccines and drugs

The prevention and treatment of pathogens using vaccines and drugs are the most robust and promising methods. Most of the vaccines for preventing waterborne infectious NTDs are undergoing clinical trial for human use; therefore, drugs are the only available option for their treatment. Vaccines in clinical trials for dengue are Dengvaxia, DENVax, LATV Δ30 [39], and for schistosomiasis are Sh-28-GST, Sm-14 and schistosome protein-calpain (Sm-p80) [40]. Pharmaceutical companies like Merck, Pfizer and GlaxoSmithKline have donated drugs to achieve the human health goals in targeting NTDs. Praziquantel is the readily available commercial drug for treating schistosomiasis, which is safe and effective to most of the people [41]. Other drugs available are albendazole, oxamniquine for *S. mansoni*, metrifonate for *S. haematobium* and artemether that kills worms in immature stage. The commonly used drugs for treating trachoma are trimethoprim-sulfonamide combinations, amoxicillin, macrolides, chloramphenicol and tetracycline, which are all effective against *Chlamydia trachomatis* [42]. The effective drugs for dengue are suramin, lovastatin, pentosan polysulfate, castanospermine (CSP) and deoxynojirimycin [43]. Dracunculiasis has no specific drug or a vaccine for its treatment and prevention [44]. Merck and WHO have collaborated for the development of ivermectin to combat against onchocerciasis (river blindness) and have certainly benefitted the African patients [45]. Merck, Eisai Co. Ltd. and GlaxoSmithKline had donated Mectizan, diethylcarbamazine and albendazole against lymphatic filariasis to the 300 million populations in endemic regions.

7. Conclusions and future perspectives

The water-borne infectious NTDs among the several other NTDs have majorly affected the poorest of the population, and with the support of health organizations and pharmaceutical industries, various steps are taken for the eradication of NTDs. To decrease the burden of NTDs, various preventive measures have to be strengthened for a sustainable solution such as maintenance of water, sanitation and hygiene in endemic areas, reducing disturbance of environment, provision of healthcare facilities to migrating populations to avoid exposure to NTDs, management of poverty in endemic area and regular risk assessment and surveillance of number of NTD cases. A decision-making body composed of scientists, microbiologists, clinicians and financial experts might be established in regions prone to water-associated infectious NTDs. The supply of water from the nearby source (stream/groundwater/piped/nonpiped) needs to be investigated to prevent the spreading of disease from few cases to endemics. The rapid point-of-care detection methods might be employed for the identification and quantification of the pathogens. With the identification of pathogen, diagnosis of severity of infection and its duration of exposure will determine the dose-dependent treatment. People sometimes choose self-medication over consulting a doctor, which might be risky as it increases the duration of illness. According to the type of water system and pathogens, proficient personnel might be required for their cleaning and removal.

Acknowledgements

The authors are grateful to the Vice Chancellor, King George's Medical University (KGMU), Lucknow, and Director, Centre for Cellular and Molecular Biology, Council of Scientific and Industrial Research (CSIR-CCMB), India, for the encouragement and support for this work. SK Saxena is also supported by CCRH, Government of India and US NIH grants: R37DA025576 and R01MH085259. The authors have no other relevant affiliations or financial involvement with any organization or entity with a financial interest in or financial conflict with the subject matter or materials discussed in the manuscript apart from those disclosed.

Author details

Shailendra K. Saxena[1,2*], Swatantra Kumar[1], Amrita Haikerwal[1] and Madan L.B. Bhatt[1]

*Address all correspondence to: shailen@kgmcindia.edu

1 Centre for Advance Research (CFAR), King George's Medical University (KGMU), Lucknow, UP, India

2 CSIR-Centre for Cellular and Molecular Biology (CCMB), Hyderabad, TS, India

References

[1] Prüss-Ustün A, Bartram J, Clasen T, Colford JM Jr, Cumming O, Curtis V, Bonjour S, Dangour AD, De France J, Fewtrell L, Freeman MC, Gordon B, Hunter PR, Johnston RB, Mathers C, Mäusezahl D, Medlicott K, Neira M, Stocks M, Wolf J, Cairncross S. Burden of disease from inadequate water, sanitation and hygiene in low- and middle-income settings: A retrospective analysis of data from 145 countries. Tropical Medicine & International Health. 2014;**19**(8):894-905

[2] Hamilton PD, Gale P, Pollard SJ. A commentary on recent water safety initiatives in the context of water utility risk management. Environment International. 2006;**32**(8):958-966

[3] Clasen T, Pruss-Ustun A, Mathers CD, Cumming O, Cairncross S, Colford JM Jr. Estimating the impact of unsafe water, sanitation and hygiene on the global burden of disease: Evolving and alternative methods. Tropical Medicine & International Health. 2014;**19**(8):884-893

[4] Brown J, Cairncross S, Ensink JH. Water, sanitation, hygiene and enteric infections in children. Archives of Disease in Childhood. 2013;**98**(8):629-634

[5] Funari E, Manganelli M, Sinisi L. Impact of climate change on waterborne diseases. Annali dell'Istituto Superiore di Sanità. 2012;**48**(4):473-487

[6] Hunter PR. Climate change and waterborne and vector-borne disease. Journal of Applied Microbiology. 2003;**94**(Suppl):37S-46S

[7] Bertuzzo E, Mari L. Hydrology, water resources and the epidemiology of water-related diseases. Advances in Water Resources. 2017;**108**:329-331

[8] Savage GB, Macintyre AK, Wicken JH, Velleman Y, Sarah V. Why water, sanitation and hygiene matter. Community Eye Health. 2013;**26**(82):27

[9] White GF, Bradley DJ, White AU. Drawers of water: Domestic water use in East Africa. 1972. Bulletin of the World Health Organization. 2002;**80**(1):63-73 discussion 61-2

[10] Ashbolt NJ. Microbial contamination of drinking water and disease outcomes in developing regions. Toxicology. 2004;**198**(1-3):229-238

[11] Leclerc H, Schwartzbrod L, Dei-Cas E. Microbial agents associated with waterborne diseases. Critical Reviews in Microbiology. 2002;**28**(4):371-409

[12] Steinmann P, Keiser J, Bos R, Tanner M, Utzinger J. Schistosomiasis and water resources development: Systematic review, meta-analysis, and estimates of people at risk. The Lancet Infectious Diseases. 2006;**6**(7):411-425

[13] Gasana J. Water and health. Air Water Borne Diseases. 2014;**3**:e129

[14] Weaver SC, Barrett AD. Transmission cycles, host range, evolution and emergence of arboviral disease. Nature Reviews. Microbiology. 2004;**2**(10):789-801

[15] Braks MAH, de RodaHusman AM. Dimensions of effects of climate change on water-transmitted infectious diseases. Air and Water Borne Diseases. 2013;**2**:109

[16] Akanda AS, Hossain F. The climate-water health nexus in emerging megacities. Eos Trans. 2012;**93**(37):353-354

[17] Alderman K, Turner LR, Tong S. Floods and human health: A systematic review. Environment International. 2012;**47**:37-47

[18] CDC. Water, Sanitation & Environmentally-Related Hygiene [Internet]. Available from: https://www.cdc.gov/healthywater/hygiene/disease/trachoma.html#one

[19] WHO Fact sheet. Schistosomiasis [Internet]. Available from: http://www.who.int/mediacentre/factsheets/fs115/en/ [Accessed: Nov 13, 2017]

[20] WHO. Human African Trypanosomiasis, Epidemiological Situation [Internet]. Available from: http://www.who.int/trypanosomiasis_african/country/en/ [Accessed: Nov 13, 2017]

[21] WHO Facts sheet. Dengue and Severe Dengue [Internet]. Available from: http://www.who.int/mediacentre/factsheets/fs117/en/ [Accessed: Jan 8, 2018]

[22] WHO Facts sheet. Leptospirosis [Internet]. Available from: http://www.searo.who.int/about/administration_structure/cds/CDS_leptospirosis-Fact_Sheet.pdf

[23] WHO Facts sheet. Lymphatic Filariasis [Internet]. Available from: http://www.who.int/mediacentre/factsheets/fs102/en/ [Accessed: Jan 8, 2018]

[24] Kappagoda S, Ioannidis JP. Prevention and control of neglected tropical diseases: Overview of randomized trials, systematic reviews and meta-analyses. Bulletin of the World Health Organization. 2014;**92**(5):356-366

[25] Geneva: World Health Organization. Working to Overcome the Global Impact of Neglected Tropical Diseases [Internet]. 2010. Available from: http://www.who.int/neglected_diseases/2010report

[26] World Health Organization. Global Status Report on Water Safety Plans: A Review of Proactive Risk Assessment and Risk Management Practices to Ensure the Safety of Drinking-Water. 2017. Available from: http://www.who.int/water_sanitation_health/publications/global-status-report-on-water-safety-plans/en/ [Accessed: Dec 15, 2017]

[27] Water Aid. WASH: The Silent Weapon against NTDs. London. 2012. Available from: https://washmatters.wateraid.org/sites/g/files/jkxoof256/files/WASH%20the%20silent%20weapon%20against%20NTDs.pdf [Accessed: Dec 15, 2017]

[28] World Health Organization. UN-Water Global Analysis and Assessment of Sanitation and Drinking-Water (GLAAS) Strategy 2017-2020. 2017. Available from: http://www.who.int/water_sanitation_health/monitoring/investments/glaas/en/ [Accessed: Dec 15, 2017]

[29] World Health Organization. Prevention of Blindness and Visual Impairment, Trachoma. 2017. Available from: http://www.who.int/blindness/causes/trachoma/en/ [Accessed: Dec 30, 2017]

[30] Kuper H, Solomon AW, Buchan J, Zondervan M, Foster A, Mabey D. A critical review of the SAFE strategy for the prevention of blinding trachoma. The Lancet Infectious Diseases. 2003;**3**(6):372-381

[31] World Health Organization, UNICEF. Prevention and Control of Schistosomiasis and Soil Transmitted Helminthiasis. Geneva: World Health Organization. 2004.

[32] Children without Worms. A comprehensive strategy for STH control. 2017. Available from: http://www.childrenwithoutworms.org/news/building-strong-foundationscww%E2%80%99s-first-eight-years-and-way-forward [Accessed: Dec 30, 2017]

[33] World Health Assembly. Elimination of Schistosomiasis. 2012. Available from: http://apps.who.int/gb/ebwha/pdf_files/EB130/B130_R9-en.pdf [Accessed: Dec 30, 2017]

[34] Global NTD Programs. Centers for Disease Control and Prevention. 2017. Available from: https://www.cdc.gov/globalhealth/ntd/global_program.html [Accessed: Dec 30, 2017]

[35] Keating J, Yujich JO, Mollenkopf S, Tediosi F. Lymphatic filariasis and onchocerciasis prevention, treatment, and control costs across diverse settings: A systematic review. Acta Tropica. 2014;**135**:86-95

[36] World Health Organization. Emergency Treatment of Drinking-Water at the Point of Use. 2013. Available from: www.who.int/water_sanitation_health/publications/.../tn5_treatment_water_en.pdf [Accessed: Nov 7, 2017]

[37] Haikerwal A, Bhatt MLB, Saxena SK. Reducing the global burden of dengue: Steps toward preventive methods. Archives of Preventive Medicine. 2017;**2**(1):028-033

[38] World Health Organization. Technical Options for Excreta Disposal in Emergencies. 2013. Available from: http://www.who.int/water_sanitation_health/emergencies/WHO_TN_14_Technical_options_for_excreta_disposal.pdf [Accessed: December 30, 2017]

[39] Saxena SK, Haikerwal A, Gadugu S, Bhatt MLB. Complementary and alternative medicine in alliance with conventional medicine for dengue therapeutics and prevention. Future Virology. 2017;**121**(8):399-402

[40] Siddiqui AA, Siddiqui AB, Ganley-Leal L. Schistosomiasis vaccines. Human Vaccines. 2011;**7**(11):1192-1197

[41] Fenwick A, Savioli L, Engels D, Robert Bergquist N, Todd MH. Drugs for the control of parasitic diseases: Current status and development in schistosomiasis. Trends in Parasitology. 2003;**19**(11):509-515

[42] Chidambaram JD, Bird M, Schiedler V, Fry AM, Porco T, Bhatta RC, Jha H, Chaudary JS, Gaynor B, Yi E, Whitcher JP, Osaki-Holm S, Lietman TM. Trachoma decline and widespread use of antimicrobial drugs. Emerging Infectious Diseases. 2004;**10**(11):1895-1899

[43] Haikerwal A, Kumar S, Kant R, Saxena SK. Potential therapeutics for dengue virus infection. Annals of Pharmacology and Pharmaceutics. 2017;**2**(8):1046

[44] CDC. Management of Guinea Worm Disease (GWD). 2017. Available from: https://www.cdc.gov/parasites/guineaworm/treatment.html [Accessed: December 30, 2017]

[45] WHO. Eliminating River Blindness. 2007. Available from: http://www.who.int/tdr/publications/documents/elimin_riverblind.pdf [Accessed: December 30, 2017]

Chapter 2

Water Quality Control in the Water Supply System for the Purpose of Preventing Legionnaires' Disease

Anita Rakić

Additional information is available at the end of the chapter

http://dx.doi.org/10.5772/intechopen.74018

Abstract

The contamination of drinking water presents an important public health concern throughout the world. Microbial hazards make the largest contribution to waterborne disease in developed and developing countries. *Legionella* bacterium, the fundamental agent of Legionnaires' disease, is a water-based organism that causes infection when inhaled in an aerosol form. Main factors influencing the survival of *Legionella* spp. are: physical and chemical properties of the water (pH value, mineral content, and presence of heavy metals), materials used in the system, system design, furring, corrosion, and dead-ends. These bacteria are transmitted during the showering by inhalation of contaminated water droplets and the highest risk of infection with a subsequent death is encountered among immunocompromised and elderly water consumers. The control of the disease has been one of the major problems in countries with low- and middle-income economies, including Croatia. The most frequently used approach to disinfect the system is a daily increase of water temperature in the heating coil. However, due to the economic issues, the residents frequently request house managers to decrease the temperature of hot water systems leading to an increased system contamination and subsequent human infections.

Keywords: *Legionella* spp, drinking water distribution system, risk assessment

1. Introduction

Environmental component research (food, water, air, and waste…) is a part of public health preventive activities with the aim of preserving the health of the population [1]. The quality as well as the quantity of water in the public supply is important for public health. Care for the health of people through environmental monitoring is a state`s obligation and should, when it comes to public health care, be funded from the budget, which will enable health professionals to work independently [2]. Water is the most valuable natural substance, which is being necessary for the survival of all living organisms. The quality as well as the quantity of water in the public supply is important for public health. Fouling is the undesired deposition of material on surface drinking water within distribution systems.

- Inorganic fouling (precipitation of inorganic crystals), "Scaling".
- Organic fouling (deposition of fat, oil, and protein etc.)
- Particle fouling (deposition of silt, clay, and humic particles etc.)

Biofouling can be formed in the water supply system during different stages of preparation and distribution of drinking water on surfaces where the water is touched by hard surfaces. Biofouling is the undesired deposition and growth of microorganisms on surfaces and particles, which can multiply on the expense of nutrients [3]. Biofouling is a complex community of autotrophic and heterotrophic microorganisms, including detritus, present on different types of substrata in all aquatic environments. The most common microorganisms present in biofouling are: *Campylobacter* spp., *Legionella* spp., *Cryptosporidium* spp., and *H. pylori*.

In aquatic ecosystems, biofouling plays an important role regarding the primary production of biofilm and its growth at the expense of nutrients from the water supply system. The growth of biofouling depends on several factors, including nutrients, hydrodynamic conditions, and sediment accumulation. Thus, nutrients have to be considered as a potential biomass for biofilm. It is an excellent indicator of ecological changes, and it increases habitat availability [4]. Also, biofouling presents a serious problem for cooling cycle in energy industry:

- Heat transfer ↓.
- Drag resistance ↑.
- Biocorrosion [3].

2. Water contamination inside the water supply system

During the operation of the water supply system, water losses, pressure or flow reduction, and water quality deterioration may be caused by increasing concentrations of various organic, inorganic, or organometallic compounds and contaminants with various microorganisms. All of these changes may be due to the interaction between water and water in the tubular and reinforced wall elements and due to the various physical, chemical, and biological reactions in the water itself during its journey from the water supply system to the consumer. A disinfectant is typically added at the end of water treatment to give a disinfectant residual to provide some protection against microbial growth and limit the effects of contamination while the water is being conveyed through the distribution system. Changes in water quality may be of greater or lesser magnitude, for example, formation of deposits or sludge, which contributes to poor smell and taste of water, and therefore, the questionable health of drinking water [5, 6]. These phenomena depend on various factors: on deposits of sediments on the walls (various oxide and oxyhydroxide products of corrosion, water solids, and organic sediments), water flow, water age, water usage dynamics, water temperature, pH, and water hardness disinfectant etc. [7]. Inside the pipeline, over time, different deposits and corrosion products are produced. At these phenomena, the highest influence is given to temperature, pH, hardness, and other chemical nutrients of water, as well as the construction material of the pipeline that distributes water to the consumer [8]. Therefore, the water supply system should be viewed as a unique biocenosis, a very complex chemical-biological reactor in which a whole series of interrelated reactions exist **Figure 1 (a)** and **(b)**.

Water pipe and the flow of water within can be considered as a bio-chemical reactor as depicted in **Figure 1**. The presence of corrosion in the water supply system increases the available surface for the colonization of microorganisms and the creation of biofilms. For the purpose of providing microbial quality, the design and operation of water supply network should prevent introduction of contaminants; disinfectant residual concentrations should be kept within a locally predetermined range, and the transit time (or age of the water after leaving the treatment plant) should be minimized. Many of the above factors also influence the biological stability of water primarily due to the formation of biofilm in which *Legionella* species (*Legionella* spp.) can be found under certain conditions.

Therefore, the most common causes of drinking water contamination in the water supply system are:

- corrosion of metal pipes and fittings [10]
- precipitation of biofilms on the inner walls of water pipes [11].

2.1. Corrosion in the drinking water distribution system

Corrosion within the water distribution system can cause water leakage, capacity loss, and deterioration of the chemical and microbiological quality of drinking water. Corrosion represents the process of unintentional destruction of construction material by physical, chemical, and

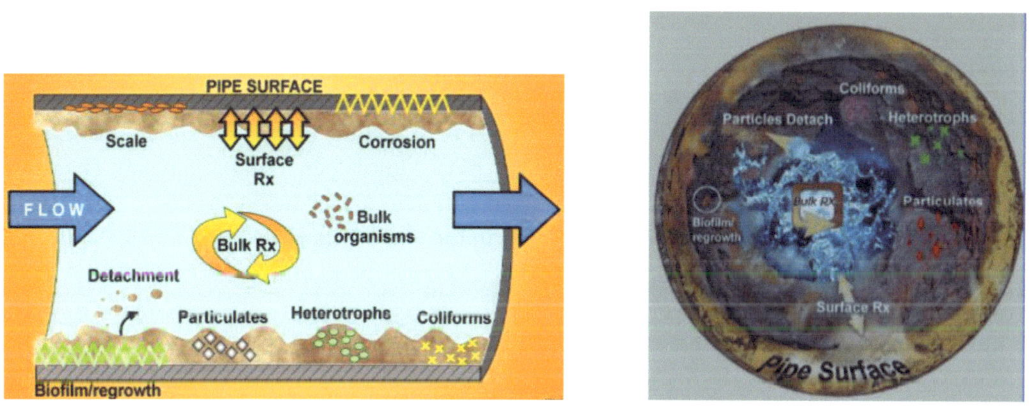

(a) b)

Figure 1. Schematic diagram of longitudinal (a) and cross section (b) water pipe as chemical-biological reactor [9].

biological environmental action, changing the structure of the material from the surface to the interior [12].

The corrosion process, metal parts within the water supply system, is influenced by a number of external and internal factors:

– **the type of metal** – less noble metals are easier to oxidize in water.

– **pH value of water** – metals and their oxides are easier to dissolve in acidic water.

– **buffer capacity** – if the relationship between H_2CO_3 and Ca is not stoichiometric, lower buffer capacity and stronger metal dissolution occur.

– **water flow** – at a higher flow, there are less concentrations of metal ions in water.

– **water temperature** – higher the temperature faster the corrosion reaction.

– **oxygen concentration** – dissolved oxygen in water is one of the most aggravating factors of corrosion.

– **electrical conductivity** – presents a danger to various metal joints (galvanized tubes) representing potential corrosion sites [13, 14].

Also, certain types of bacteria (sulfate-reducing bacteria) in the aqueous medium affect the changes in metal tube construction materials, thus encouraging and enhancing corrosion (biocorrosion) of metal surfaces. Microbially induced corrosion causes the displacement of the corrosion system with a two-component mechanism (metal-medium) into a corrosion system with a three-component mechanism (metal-medium-biofilm) [15].

2.1.1. The presence of heavy metal ions in the drinking water

The corrosion of metal pipes increases the concentration of the heavy metal ions having harmful effects on human health. It has been found that important factors for the occurrence of *Legionella* spp. are the drinking water distribution system and the corrosion of pipes, pumps, valves, other appurtenances, and cooling towers [16]. It has been proven that some metal ions retard while others have a biostimulating effect on the growths of *Legionella* spp. Heavy metals include metal groups having a relative density greater than 5.0 g/cm^3. In the atmosphere, water and soil fall short of a variety of natural sources, mostly due to urbanization and industrial processes. Water is deposited at the bottom of the water surface as a hard soluble carbonate, sulfate, or sulfide. They are not biodegradable and have the ability to bioaccumulate in living organisms. Drinking water may come to an end if water sources are contaminated with heavy metals, but the main source of corrosion products are metal structures in the water supply system [17, 18].

Heavy metals falling into human organisms can lead to:

- blocking of basic biological functional groups of biomolecules (e.g., proteins and enzymes).
- the displacement of essential metal ions (Fe, Cu, and Zn).
- modification of active forms of biomolecules.

http://dx.doi.org/10.5772/intechopen.74018

2.2. Creating deposits in the drinking water distribution system

Deposit represents the mineral deposit, which consists primarily of calcium and magnesium carbonate. In pipes in urban water supply systems, lime deposition occurs, primarily on the surfaces of the heat transfer system, and this is particularly evident in water distribution pipelines causing water hardness. The most responsible for the deposit formation in the hot water supply systems is the transitory deposit. Namely, by heating the water, the soluble bicarbonates convert into the hard soluble carbonates by reactions:

$$Ca^{2+}{}_{(aq)} + 2HCO_3^{-}{}_{(aq)} \rightarrow CaCO_{3\,(s)} + H_2O_{(l)} + CO_{2\,(q)} \quad (1)$$

$$Mg^{2+}{}_{(aq)} + 2HCO_3^{-}{}_{(aq)} \rightarrow MgCO_{3\,(s)} + H_2O_{(l)} + CO_{2\,(q)} \quad (2)$$

Due to deposition of the deposit, that is, hard soluble calcium and magnesium carbonate and release of CO_2 (Eqs. 1 and 2), the corrosive action of water is increased, particularly above 60°C. Deposits can cause many problems in heating systems and in the distribution network. Its deposition on heat exchanger walls reduces heat transfer and water flow, which can lead to clogging of certain parts of the system. The heating efficiency can be reduced by up to 2–6%, which means increased heating costs and higher CO_2 emissions. Because of the unequal deposition of the lime over the water pipes, local overheating, water vapor, and difficulties in the operation of boiler plants (creating noise in the system) occur. Due to the heat released during the operation of the pump, the deposition of the deposits and the inside of the pump housing can occur, which reduces the flow of water. The porous structure of the deposit favors the propagation of microorganisms by protecting them from the influence of disinfectants and the influence of hot water. Also, water heating increases electrical conductivity and galvanic corrosion [19].

2.3. Creating biofilms in the drinking water distribution system

The water supply system is made of the following main groups of objects: source water, treatment, storage, and distribution system. *Legionella* spp. has been shown to be harbored within biofilms formed within different parts of drinking water supply [20].

In the water supply system, biofilms can be created in different stages of preparation and distribution of drinking water on surfaces where the water is touched on a solid substrate (**Figure 2**). For example, locations for biofilm growth in drinking water systems are:

- **inner walls of wells**, well plumbing, plumbing pump, etc.
- **treatment for waters**: surfaces of filter media (sand, activated carbon), membranes, etc.
- **drinking water distribution**: inner walls of pipes made of mineral, metallic and polymer surfaces, hoses, etc.
- **drinking water reservoirs**: walls, floors, ceilings.

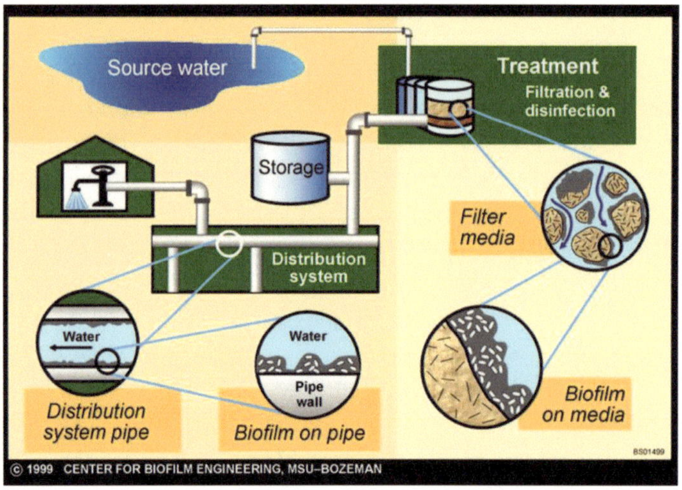

Figure 2. Scheme drinking water distribution system [21].

The risk of significant surge, and hence water quality problems, is greater in long unbranched pipes than in branched pipes, because branched pipes reduce surge. Microbial growth in water depends on temperature, nutrient content, and disinfectant concentration. In a network, it will also depend on the composition of the internal pipe surfaces, but this effect cannot be predicted. It is relatively easy to predict the temperature and nutrient content in the mixed water, as they are derived from the flow-weighted average of the values in the constituent waters [22]. Disinfectant concentration is dependent on the degree of decay that the constituent waters contain up to the point of mixing and the type of disinfectant, which was used in the constituent waters, the blending proportions and the chemical reactions, which occur between the disinfectant species [23]. Environmental conditions in the aqueous fluid, which are favorable for growth (e.g., which contain sufficient substrate and growth nutrients) of the attached cells will lead to the growth, division, and formation of new cells and produce a matrix of extracellular polymeric substances (EPS), which adhere to each other and to the surface (substratum). Such accumulation of cells and EPS, along with any trapped inert particles and organic matter, is termed "the biofilm". Products of cell metabolism and biotransformation are passed back into the aqueous phase together with the cells that become detached from the biofilm [4]. Biofilms as microorganism reservoirs depend on nutrients from the water. Hygienically relevant bacteria detected in drinking water biofilms were as follows: *Campylobacter jejuni, Campylobacter coli, Faecal streptococci, Escherichia coli, Helicobacter pylori, Legionella pneumophila, Mycobacterium avium,* and *Pseudomonas aeruginosa* [4]. Low temperature and high disinfectant levels generally inhibit microbial growth, which otherwise depends on nutrient level. Thus, a particular water may exhibit low microbial growth at low temperature and low disinfectant residual, as may another water at a higher temperature with a higher residual, but a mix of the two may support high microbial growth [24]. In **Figure 3**, the biofilm formation within the plumbing tube is shown schematically.

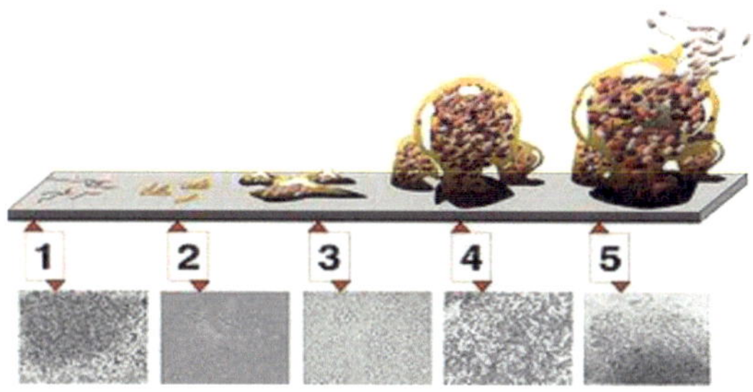

Figure 3. Representation of biofilm formation [25].

Biofilm formation on solid surfaces involves multiple stages (**Figure 3**) and five steps in biofilm formation are:

1. In the environment free-living microorganisms – Primary adhesion, reversible, and irreversible.
2. Formation of microcolonies, Irreversible attachment.
3. Development of a continuous biofilm, Maturation I.
4. Sloughing off of biofilm parts, Maturation II.
5. Transport of biofilm particles (flocs) throughout the system, initiation of further biofilm formation, Dispersion.

2.3.1. *Influence of environmental factors on the development of biofilms in the pipeline*

Environmental conditions and sites that favor the emergence of biofilms and the reproduction of different species and numbers of microorganisms in water supply systems are the subject of many scientific researches [26, 27]. Changes may be the result of permanent or intermittent influences of physicochemical factors of the water environment [28, 29], which can be directly influenced by a particular microbiological population or indirectly through interactions between members of the microbiological community. While being transported through a pipeline network, treated water gets in contact with a number of different surfaces. Therefore, no materials to which the drinking water is exposed in the network should promote microbial growth or discharge any contaminants, which can support microbial growth into the water. The degree to which microorganisms attach to the wall of the tubes, their growth, and reproduction depend on environmental conditions [30–32].

The rate of biofilm growth depends on:

- the physicochemical properties of water (temperature, pH, hardness, organic materials, nutrients, disinfection residual concentrations, and heavy metals), water flow velocity, andcorrosion of distribution system pipes and fittings.

2.3.1.1. Impact of the surface roughness of the pipe inside the water supply system on biofilm development

The roughness of the surface of the water piping material is essential for the development of biofilms within them [33]. The roughness of the pipe surface can be a result of the corrosion process of the piping material, thereby enhancing the colonization of microorganisms and the creation of biofilms on such surfaces [34, 35]. Materials used to make pipes within a water supply network should have: a small roughness of walls and joints; resistance to internal and external pressures; resistance to corrosion and the impact of aggressive groundwater; resistance to decaying currents; and waterproofing inwards and outwards. Smaller concentration of microorganisms on the surfaces of smooth tubes (plastic materials) was found in comparison to rugged surfaces (cast iron, copper, and galvanized tubes) [32, 35]. Plastic materials are resistant to corrosion and the concentration of heavy metals in drinking water is excluded due to corrosion [36, 37]. Their surface roughness is lower than other materials, and since the tubes do not corrode, the inside of the pipe surface remains smooth, and has a longer life span.

2.3.1.2. Influence of chemical composition, nutrient content, and disinfectant of these water temperatures on the development of biofilms

The presence and concentration of nutrients, oxygen concentration, and optimal temperature and pH within biofilms provide support for growth and propagation of different microorganisms of heterogeneous populations. Bacterial biofilms change their properties depending on the concentration and composition of the nutrients in the water, which is mutated, thus increasing their ability to survive in the most difficult conditions. Microorganisms present in biofilms often develop increased tolerance to biocides [38–40]. The resulting biofilms are difficult to remove by disinfection, especially with the hardest available surfaces (pipe edges, T – profiles of water pipes, rough surfaces in water pipes) or the surface where water retention occurs.

2.3.1.3. Hydrodynamic conditions in the water supply system

Drinking water supplies from water to the point of consumption depends on topography and a mixture of gravity and pressure pipes is often used in the process. The purpose of a system of pipes is to supply water at adequate pressure and flow.

The conditions of the laminar flow of flow media are always present in the water supply network. In the laminar flow, a boundary layer is formed along the wall of the pipe in which the flow rate is smaller and decreases with a reduction of the distance from the surface. The result is a quiet microclimate of biofilm. At higher water flows, greater shear forces and biofilm removal from the surface occur. The consequence of this is the weaker development of biofilms on the surface of water pipes. However, if the biofilms develop at a high friction in turbulent flow, they are stronger, more strongly attached to the substrate, having higher density and physiological activity than with low friction or laminar flow [41]. The increased flow rate in the pipes often causes rupture of the incrustation from the inner surface of the tubing or the partial

http://dx.doi.org/10.5772/intechopen.74018

mobilization of the deposits that mingle with the water and create additional pressure on the filtration systems leading to a reduction in the quality of drinking water. The created biofilms are difficult to remove with slow flow of water and hard-sided surfaces, and thus encourage additional microcorrosion on metal walls beneath the biofilm layer [42, 43]. Reduced flow and/or stagnation of the flow may cause uncontrollable growth of biofilms, which can lead to mechanical seizures of water pipes, greater risk of microorganism development, disruption of heat transfer, increased friction resistance and metal biocorrosion. For the purposes of maintaining microbial quality, it is important to minimize transit times and avoid low flows and pressures [44]. It is highly recommended to avoid low-flow dead-ends and loops, although it is not always possible in practice. Low-flow sections of dead-ends should be as short as possible. Problems can be caused by both dead-ends and loops in the system, as they create long residence times and sections in which sediments are likely to collect. The risk may be elevated in seasons with greater rainfall, where soil moisture conditions will increase the likelihood of a pressure gradient developing from the soil to the pipe. Water quality may also deteriorate on recharging where surges may dislodge biofilm, leading to esthetic problems. Due to the fact that intermittent systems are intrinsically vulnerable, hazards should be controlled in the immediate vicinity of pipes. In the longer term, it is important to reduce intermittence; in certain areas, this can be easily achieved by using or rehabilitating service reservoirs. A backflow event will be a sanitary problem if there is cross-connection between the potable supply and a source of contamination. Reviews of waterborne disease outbreaks in municipal systems often identify backflow events as a causative factor.

2.4. Microbiological composition of biofilms

The emerging biofilms in drinking water distribution systems can be transient or long-lived habitats for sanitary–hygienically important microorganisms. Biofilm is composed of a large number of bacteria including *Legionella* spp., *Klebsiella* spp., *Pseudomonas* spp., *Mycobacterium, E. coli,* and other organisms such as protozoa (amoebae), parasites, and enteroviruses [27, 28, 45]. These microorganisms can be linked to an already existing biofilm, where so integrated can survive for several days or weeks, depending on the biology and ecology of the organism and environmental conditions. Microorganisms in biofilm are often resistant to biocides and it is hard to remove them, especially from hard-to-access places such as pipe edges, T – joints, and rough inner surfaces. In this way, they pose a potential danger of drinking water contamination, and thus a health hazard for humans [4, 42, 43]. There are several reasons why biofilms are of interest in medical, industrial, and natural contexts [46]. For example, they sometimes act as reservoirs from which pathogens are disseminated **Table 1**.

In the food industry, the development of biofilms on machine surfaces serves as a protection against pathogenic microorganisms from disinfectants. Likewise, in many industries, biofilms are responsible for major economic losses. *Legionella* spp. is frequently found in biofilms formed within drinking water pipelines. Also, biofilms are the cause for the contamination of water systems by legionellae. Legionellae regularly colonize water supply systems and devices, and people are infected through the inhalation of a contaminated water aerosol.

Industry	Problems
Potable water distribution	Reduction of flow rates Taints Unacceptably high bacterial numbers
Structural steelwork, pipelines	Accelerated corrosion
Heating and cooling operation	Reduced efficiency
Food processing	Reservoir of spoilage and potentially pathogenic microorganisms; possible survival of pathogens through underprocessing
Fluid transfer (general)	Reduction of flow rates and blocking of pipes

Table 1. Examples of industrial problems caused by biofilm formation.

2.4.1. *Legionella spp.*

As noted previously, *Legionella* bacteria thrive in biofilms. There are numerous advantages to bacteria growing in biofilms. For example, they can act as reservoirs from which the dissemination of pathogens may occur. *Legionella* spp. has been shown to be harbored within biofilms formed within drinking water pipelines. This investigation attempted to define these metal requirements and to adjust the composition of our defined medium accordingly. Defining the metal requirements of *Legionella* spp. is an important step in understanding its metabolism and in devising special media for isolation and maintenance. These findings suggest that metal plumbing components and associated corrosion products are important factors in the survival and growth of *Legionella* spp. [47]. Microbial hazards make the largest contribution to waterborne disease in developed and developing countries. Nevertheless, chemicals in water supplies can cause serious health problems – whether the chemicals are naturally occurring or derive from sources of pollution. *Legionella* spp. are ubiquitous intracellular microorganisms, an opportunistic human pathogen whose natural environment is aquatic environment [27, 46]. Supporting growth and survival of *Legionella* spp. in the environment is enhanced by their ability to form symbiotic relationships with other larger microorganisms including protozoa (*Acanthamoeba, Hartmanella, Valkampfia, Naegleria, …*). Actually, *Legionella* spp. lives inside protozoa (amoebae), which can be found in domestic water supplies [48, 49]. Biofilms can occlude pipework, resulting in areas of poor flow and stagnation with higher risk of *Legionella* growth. Also, the presence of both biofilms and amoebae has a twofold protective effect for the bacteria in the system [50]. For example, it increases the organic load and inactivates residual chloride levels of disinfectant. Likewise, biofilms and bacteria within them (including *Legionella* spp.) grown inside amoebae are more tolerant to chlorine and other antimicrobial agents at concentrations above those commonly used to disinfect water supplies [48]. Because bacteria in biofilms are relatively resistant to standard water disinfection procedures, *Legionella* are able to enter and colonize potable water supplies [40].

Enhanced resistance against are disinfectants by adaptation to biofilm conditions, intracellular localization in cells and cysts (resting stages of protozoa), *Legionella* spp. are typical biofilm-related pathogens as they survive in protozoa, which graze on biofilms.

2.5. Legionnaires' disease

Bacteria of the genus *Legionella* are important causes of the both community-acquired and nosocomial pneumonia. Legionnaires' disease (LD) is a form of interstitial pneumonia that is normally transmitted via aerosol, for example, from the environment to humans mainly through the inhalation of contaminated aerosols (e.g., mist droplets containing the bacteria) [32]. Contaminated water sources such as domestic hot-water systems, swimming and spa pools, respiratory therapy equipment, cooling towers, fountains, and other devices that utilize public water supply can produce the aerosol containing legionella bacteria. Legionnaires' disease affects all age groups and more often elderly people, particularly those with chronic cardiac, pulmonary and renal diseases. Immunocompromised patients are particularly at high risk. Immunosuppression is commonly associated with nosocomial infection with legionella (90%) compared with community-acquired cases [51–53]. Legionnaires' disease involves multiple organs but pneumonia is the most common presentation, especially in immunocompromised patients [33]. Legionella is responsible for 2–15% of pneumonias in the general population. There are no reliable clinical, laboratory, or radiological indicators that could with certainty differentiate Legionnaires' disease from other pneumonias. Consequently, clinical doubts should be always confirmed with special laboratory tests (culture, detection of antigen in respiratory secretion or urine, serologic tests, and molecular diagnostics) [34]. LD is an important cause of hospital-acquired pneumonia [35]. The appropriate antibiotic should be administered in the treatment of Legionnaires' disease as early as possible. New macrolides, azithromycin, and fluoroquinolones, doxycycline, and rifampicin are traditionally recommended for the treatment of Legionnaires' disease [34]. Therefore, all patients with a more severe clinical picture or severe accompanying chronic diseases and compromised immunity require an antibiotic efficient against legionella during the onset of the treatment [36, 37]. In epidemiology, we divide diseases into antroponoses and zoonoses, and legionary disease is neither, because it is related to the environment as a source, so we can call it econoses. Mainly, it is related to human dwellings and conditions that man created by its activities attempting to please himself, because the occurrence of disease in natural conditions is extremely rare. It is the price of progress and punishes any omission in technical performance of building, but particularly punishes inadequate maintenance of humid air conditioning and water supply systems in buildings of public importance.

2.5.1. *Travel-associated Legionnaire's disease*

Tourism is a growing economy sector. According to the data from the World Health Organization, 940 million of people were traveling in 2010. Millions of people cross international borders every day, traveling in various parts of the world exposing them to a different health environment [54, 55]. Climate change, globalization, and other outcomes of industrialization change the epidemiology of infectious diseases, but even more noticeable, the appearance of old and forgotten diseases in areas where they have not previously occurred, with new clinical features, resistance to antimicrobial therapy, and risk for human health. Tourism, uncontrolled expansion of cities, exotic animals as pets, the uncontrolled import of food and population movements, opens the real possibility of the introduction of new pathogens. Over recent years in our country, there are a growing number of tourists, who expect, when arriving in a new environment, this issue to be monitored [56, 57]. These can be epidemics caused by contagious diseases with the multiple causal pathways. The most common are viral infections of the upper respiratory tract, colds, but poor sanitation and unregulated water supply may also be related to epidemics of acute intestinal infectious diseases, in the last few years. However, it has become less frequent due to the significant improvements in water supply and the availability of fecal isolation. While spreading of contagious diseases can be hardly prevented via general precautions, including intestinal contagious diseases particularly with the drinking water transfer route, our responsibility is to ensure proper water supply to provide sufficient quantities of potable drinking water that must be constantly chlorinated as required by sanitary regulations in all developed countries. Nevertheless, epidemic incidents occur in the form of hydraulic epidemics. As a rule, they occur at small alternative water supply systems used in the peak season when existing water supply systems do not meet the amount of water needs due to the high consumption. Weather conditions can also affect waterborne epidemics – if there should be lack of timely reaction of expert services in watercools regarding the increased chlorination of drinking water in the case of large amounts of precipitation [58]. Legionnaire's disease has public health significance because of hard clinical picture and possible lethal issue. For objects where the disease is found, it means negative reputation and special requirements for continual epidemical measures. Implementation of preventive measures in different accommodation sites is the key to control of travel-associated Legionnaires' disease, which occurs after exposure to contaminated water aerosols. Therefore, it is important to monitor these facilities and educate their staff to implement preventive measures. In hotels with year around operation, the water system is continuously rinsed thus reducing the concentration of Legionella bacteria in the system. In hotels with seasonal work, water is stagnating while the property is closed and, if preventive measures before the opening of the facility are not implemented, the risk of Legionnaires' disease is increased. All hotels before opening have to pursuit epidemical measures to decrease risk of legionary disease [58]. Sampling and analysis of water is often carried out in order to gain orientation on the potential risk of the accommodation site under surveillance. Those arrangements are persuiting by virtue decree of sanitary inspection and with expert control employees from epidemiological department. Each epidemiological method used for disease prevention and health protection of its population is of all significance and vital to providing quality and safety in the tourism sector. Tourism industry and its workers should be aware of the public health safety importance being an inevitable requirement for the improved service quality ensuring high-quality guest experiences as well as competitive advantage of tourism industry.

2.5.2. Health risks associated with potable water on ships

Waterborne outbreaks have been associated with bunkering water of poor quality and causes such as:

- contaminated water supplied at the port.
- contaminated bunkered water.
- cross-connections between potable and non-potable water.

- poor design and construction of potable water storage tanks.
- unadequate disinfection.

In those ports which do not have a safe source of water, contaminated water bunkered from port can lead to a number of outbreaks due to enterotoxigenic *Escherichia coli, Giardia lamblia,* and *Cryptosporidium.* Legionnaires' disease is one of the most widely known forms of legionellosis. It is a type of pneumonia one acquires by inhaling aerosols, which contain an excessive quantity of *Legionella* bacteria. There are several reasons why ships are considered high-risk environments for the proliferation of *Legionella* spp. [46]. Firstly, source water quality can pose a health concern if it is treated only with a residual disinfectant prior to or upon bunkering, or untreated at all. Secondly, water storage and distribution systems on ships are complex, potentially providing numerous opportunities for bacterial contamination, as the risk of surge and back-siphonage increases due to ship movement [59]. Thirdly, there might be considerable variations in the temperature of potable water (caused, for example, by high temperatures in the engine room). High-water temperatures characteristic of some tropical regions can enhance the risks of bacterial growth and occurrence of *Legionella* contamination in cold water systems. Finally, proliferation is further stimulated by long-term storage and stagnation in tanks or pipes. It should be noted that *Legionella* spp. can proliferate in warm-water temperatures (ranging between 25 and 50°C), such as those encountered in showerheads and pools, which can lead to exposure through aerosolization occurring in showers and other plumbing fixtures. The production of water on ships can sometimes yield its own potential health problems. There are several different processes by means of which water can be produced on ships, such as reverse osmosis or seawater evaporation. Seawater gets demineralized through desalination, thus becoming more corrosive and capable of shortening the life of containers and conduits. Desalinated water can also cause health issues associated with a lack of minerals in seafarers' diets or the consumption of dissolved metals (e.g., lead, nickel, iron, cadmium, or copper) from corrosion products. Also, passengers and crew often characterize desalinated water as bland, flavorless and therefore unacceptable. Ships' evaporation systems are supplied with seawater, which is usually led directly into the evaporator after having been sucked in through sea chests. Reverse osmosis involves pretreatment and transport of water across membranes under pressure so that salts are excluded.

2.5.2.1. Drinking water supply and transfer chain on ships

Generally, the ship drinking-water supply and transfer chain consists of three major components:

1. the source of water coming into the port.
2. the water transfer and delivery system (including hydrants, hoses, water boats, and water barges), which provides numerous opportunities for the ingress of contaminants into the drinking water
3. the ship water system, which includes storage, distribution and onboard production of drinking water from overboard sources, such as seawater

To obtain reliable and comparable information about the sanitary status of the potable water installation, it is recommended that samples be taken at the same places (e.g., always at the tank and from the bridge deck) [60].

3. Conclusions

The presence of *Legionella* spp. in water is considered to pose a health risk for consumers, therefore monitoring and determination of risk factors on *Legionella* spp. presence is important for water quality preservation. The aim is to significantly improve health and well-being of populations, as well as to reduce health inequities and to ensure sustainable people-centered health systems [61–63].

In order to prevent the spread of epidemics, the European Working Group for Legionella Infections (EWGLI) was established. In the hotel accommodation, the most frequent preventive measure to reduce the risk of Legionnaires' is the implementation of pasteurization. The scientific approach proved the dependence of *Legionella* spp. presence with the concentration of metals and pointed to the potential public health problem. Water supply organizations should adopt network design and operating strategies that prioritize issues closely linked to water supply hygiene.

3.1. Monitoring: the key to successful antifouling strategies

Monitoring is crucial for timely detection of biofilms and countermeasure optimization. Because, technical systems are not sterile, carry biofilms (cleaning is more important than killing biofilm organisms), nutrients are potential biomass (nutrient limitation), disinfection is not cleaning and biofilm management: keeping biofilm formation below threshold of interference. Too, it is therefore important to carry out regularly preventive measures to reduce the risk of Legionnaires' disease for tourists and facility staff.

It is also important to list some of the recommendations for prevention:

- identify and prevent low pressures (particularly negative pressures) in the system.
- prevent pressure surges in the network and use low capacity hot water tanks.
- network design, which minimizes the risks of contamination during operational activities and avoids water stagnation, prevent cross-connections and backflow.
- design and operate service reservoirs to avoid stagnation and contamination by ingress.
- select construction materials that do not promote microbial growth.
- constructions or pipelines with less curvature, dead zones,
- body installation material – polymeric or copper-based pipes.

- use softened water for depositing the pipes, water must not be excessively hard (Ca^{2+} and Mg^{2+}).
- it is necessary to maintain hot water temperatures above 50°C and a cold water temperature below 20°C.
- use of disinfectants that penetrate biofilms, to limit the risk of re-contamination.
- shock treatment often inadequate (hyperchlorination, heat treatment, hydrogen peroxide plus peracetic acid, etc.) to eradicate legionellae contained in biofilms.

3.2. Future prospects for the prevention of Legionnaires' disease

With the development of new technologies and the spread of new findings related to *Legionella* spp., new perspectives for the prevention of Legionnaires' disease are opening up.

Therefore, it would be beneficial to carry out projects and/or research that would be based on finding the relationship between energy savings for water heating in households, hotels, ships, and risks associated with the occurrence of Legionella spp. It would also be useful to develop mathematical models (based on risk analysis) to predict the occurrence of Legionella as well as to determine the long-term effect of currently used disinfection strategies on Legionnaires' disease in the world. Therefore, it is advisable to provide:

1. Development of methods for *Legionella* spp. enumeration in water distribution system sediments, in suspension and in protozoa.

2. Assessment of the main economic, environmental, and genetic factors associated with *Legionella* colonization and persistence in households, hotels, boats, etc. within their water distribution systems.

3. Assessment of the burden of legionellosis associated with water contamination.

4. Determination of the effect of various disinfection methods on *Legionella* persistence in pilot scale studies.

5. Development of a mathematical model for *Legionella* colonization based on monitoring and pilot scale data and validation of the mathematical model in full scale studies.

6. Development of the evidence-based risk management strategy for *Legionella* in households, hotels, boats, etc. within their water distribution systems.

It is to be expected that the results obtained will provide a fresh insight into Legionelle's ecology in hot water supply systems. They will present new information necessary for the control of *Legionella* in the in-house water distribution systems of low economic regions. Main factors affecting the multiplication and persistence of these bacteria will be established. The potential control mechanisms (optimal disinfection conditions and predictive modeling) will be established. The obtained data will be useful not only scientifically but also for epidemiological control, clinical diagnostics and household, hotel or ship management.

Acknowledgements

I would like to thank my colleagues at the Public Health Institute of Split and Dalmatia County who contributed in different ways to this article especially prof Katji Matešan.

Author details

Anita Rakić

Address all correspondence to: anita.rakic@st.t-com.hr

Public Health Institute of Split and Dalmatia County, Croatia

References

[1] The World Health Organization (WHO). Guidelines for Drinking - Water Quality, Addendum: Microbiological Agents in Drinking Water. 2nd ed. WHO Press; 2007

[2] www.euro.who.int/__data/assets/pdf_file/0007/147724/wd09E_Health2020_111332.pdf [Accessed: November 5, 2016]

[3] Moel PJ, Verberk JQ, van Dijk JC. Drinking Water Principles and Practices. World Scientific Publishing Co. Pte. Ltd; 2006. p. 29

[4] Flemming HC. Biofilm highlights. In: Flemming HC, Wingender J, Szewzyk U, editors. Springer, Verlag; 2011. pp. 81-84, 189-214

[5] Wang H, Chun H, Zhang L, Li X, Zhang Y, Yang M. Effects of microbial redox cycling of iron on cast iron pipe corrosion in drinking water distribution systems. Water Research. 2014;**65**:362-370

[6] Enning D, Garrelfs J. Corrosion of iron by sulfate-reducing bacteria: New views of an old problem. Applied and Environmental Microbiology. 2014;**80**(4):1226-1236

[7] Chu C, Lu C, Lee C. Effects of inorganic nutrients on the regrowth of heterotrophic bacteria in drinking water distribution systems. Journal of Environmental Management. 2005;**74**:255-263

[8] Bédard E, Fey S, Charron D, Lalancette C, Cantin P, Dolcé P, Laferrière C, Déziel E, Prévost M. Temperature diagnostic to identify high risk areas and optimize *Legionella pneumophila* surveillance in hot water distribution systems. Water Research. 2015;**71**:244-256

[9] The United States Environmental Protection Agency (EPA). Water Distribution System Analysis: Field Studies, Modelling and Management, A Reference Guide for Utilities. EPA; 2008. p. 31

[10] Little BJ, Gerke TL, Lee JS. Mini-review: The morphology, mineralogy and microbiology of accumulated iron corrosion products. Biofouling. 2014;**30**(8):941-948

[11] Wang H, Chun H, Li X. Characterization of biofilm bacterial communities and cast iron corrosion in bench-scale reactors with chloraminated drinking water. Engineering Failure Analysis. 2015;**57**:423-433

[12] Videla HA. Microbially induced corrosion: An updated overview. International Biodeterioration and Biodegradation. 2001;**48**:176-201

[13] Bargellini A, Marchesi I, Righi E, Ferrari A, Cecetti S, Borella P, Rovesti S. Parameters predictive of *Legionella* contamination in hot water systems: Association with trace elements and heterotrophic plate counts. Water Research. 2011;**45**(6):2315-2321

[14] Buse HY, Schoen ME, Ashbolt NJ. Legionellae in engineered systems and use of quantitative microbial risk assessment to predict exposure. Water Research. 2012;**46**:921-933

[15] Yang F, Shi B, Bai Y, Sun H, Lytle DA, Wang D. Effect of sulfate on the transformation of corrosion scale composition and bacterial community in cast iron water distribution pipes. Water Research. 2014;**1, 59**:46-57

[16] Eboigbodin KE, Seth A, Biggs CA. A review of biofilms in domestic plumbing. Journal of American Water Works Association. 2008;**100**:131-138

[17] Swietlik J, Raczyk-Stanisławiak U, Piszora P, Nawrocki J. Corrosion in drinking water pipes: The importance of green rusts. Water Research. 2012;**46**(1):1-10

[18] Graube F, Grahl S, Rostkowski S, Beckmann M. Optimisation of water-cannon cleaning for deposit removal on water walls inside waste incinerators. Waste Management & Research. 2016;**34**(2):139-147

[19] Borella P, Montagna MT, Romano-Spica V, Stampi S, Stancanelli G, Triassi M, Neglia R, Marchesi I, Fantuzzi G, Tatò D, Napoli C, Quaranta G, Laurenti P, Leoni E, De Luca G, Ossi C, Moro M, D'Alcalà GR. Risk factors associated with isolation of *Legionellae* from domestic hot water. Emerging Infectious Diseases. 2004;**10**:457-464

[20] Wingender J, Flemming HC. Biofilms in drinking water and their role as reservoir for pathogens. International Journal of Hygiene and Environmental Health. 2011;**214**:417-423

[21] http://www.biofilm.montana.edu/resources/images/industrial-systems-processes/water-distribution-system.html [Accessed: July 21, 2014]

[22] Bédard E, Fey S, Charron D, Lalancette C, Cantin P, Dolcé P, et al. Temperature diagnostic to identify high risk areas and optimize *Legionella pneumophila* surveillance in hot water distribution systems. Water Research. 2015;**71**:244-256

[23] Wang H, Edwards M, Falkinham JO, Pruden A. Molecular survey of the occurrence of *Legionella* spp., *Mycobacterium* spp., *Pseudomonas aeruginosa,* and amoeba hosts in two chloraminated drinking water distribution systems. Applied and Environmental Microbiology. 2012;**78**(17):6285 6294

[24] Allegra S, Grattard F, Girardot F, Riffard S, Pozzetto B, Berthelot P. Longitudinal evaluation of the efficacy of heat treatment procedures against *Legionella* spp. in hospital water systems by using a flow cytometric assay. Applied and Environmental Microbiology. 2011;**77**(4):1268-1275

[25] https://en.wikipedia.org/wiki/Biofilm#/media/File:Biofilm.jpg [Accessed: 2017-2112-12]

[26] Moritz MM, Flemming HC, Wingender J. Integration of *Pseudomonas aeruginosa* and *Legionella pneumophila* in drinking water biofilms grown on domestic plumbing materials. International Journal of Hygiene and Environmental Health. 2010;**213**:190-197

[27] Scheikl U, Sommer R, Kirschner A, Rameder A, Schrammel B, Zweimüller I, et al. Free-living amoebae (FLA) co-occurring with legionellae in industrial waters. European Journal of Protistology. 2014;**50**(4):422-429

[28] Taylor M, Ross K, Bentham R. *Legionella, Protozoa* and Biofilms: Interactions Within Complex Microbial Systems. Microbial Ecology. 2009;**58**(3):538-547

[29] Rakić A, Perić J, Influence FL. of temperature, chlorine residual and heavy metals on the presence of *Legionella pneumophila* in hot water distribution systems. Annals of Agricultural and Environmental Medicine. 2012;**19**:431-436

[30] Farhat M, Trouilhé MC, Foret C, Hater W, Moletta-Denat M, Robine E, Frère J. Chemical disinfection of *Legionella* in hot water systems biofilm: a pilot- scale 1 study. Water Science and Technology. 2011;**64**(3):708-714

[31] Ditommaso S, Giacomuzzi M, Gentile M, Moiraghi AR, Zotti CM. Effective environmental sampling strategies for monitoring *Legionella* spp. contamination in hot water systems. American Journal of Infection Control. 2010;**38**(5):344-349

[32] Burillo A, Pedro-Botet ML, Bouza E. Microbiology and Epidemiology of Legionnaire's Disease. Infectious Disease Clinics of North America. 2017;**31**(1):7-27

[33] Han Htwe T, Khardori NM. Legionnaire’s Disease and Immunosuppressive Drugs. Infectious Disease Clinics of North America. 2017;**31**(1):29-42

[34] Kuzman I, Puljiz I. Legionarska bolest: praktični pristup dijagnostici i liječenju. Medicus. 2005;**14**(1):107-113

[35] Sabria M, Yu VL. Hospital-acquired legionellosis: solutions for a preventable infection. The Lancet Infectious Diseases. 2002;**2**(6):368-373

[36] Roig J, Rello J. Legionnaires' disease: A rational approach to therapy. The Journal of Antimicrobial Chemotherapy. 2003;**51**:1110-1129

[37] Rapp RP. Pharmacokinetics and pharmacodynamics of intravenous and oral azithromycin: Enhanced tissue activity and minimal drug interactions. The Annals of Pharmacotherapy. 1998;**32**:785-793

[38] van der Kooij D, Veenendaal HR, Scheffer WJ. Biofilm formation and multiplication of Legionella in a model warm water system with pipes of copper, stainless steel and crosslinked polyethylene. Water Research 2005;**39**:2789-2798

[39] Serrano-Suárez A, Dellundé J, Salvadó H, Cervero-Aragó S, Méndez J, Canals O. Microbial and physicochemical parameters associated with *Legionella* contamination in hot water recirculation systems. ESPR. 2013;**20**:5534-5544

[40] Nguyen C, Elfland C, Edwards MA. Impact of advanced water conservation features and new copper pipe on rapid chloramine decay and microbial regrowth. Water Research. 2012;**46**(3):611-621

[41] Obrdalj M. Gubici vode u vodoopskrbnim sustavima. Hrvatska vodoprivreda. 2011;**1**:17-19

[42] Ohl L, Horn H, Behaviour HD. of biofilm systems under varying hydrodynamic conditions. Water Science and Technology. 2004;**49**(11–12):345-351

[43] Liu Z, Lin YE, Stout JE, Hwang CC, Vidic RD, Yu VL. Effect of flow regimes on the presence of Legionella within the biofilm of a model plumbing system. Journal of Applied Microbiology. 2006;**101**(2):437-442

[44] Shang F, Uber JG, Rossman LA. Modeling reaction and transport of multiple species in water distribution systems. Environmental Science & Technology. 2008;**42**(3):808-814

[45] Marchesi I, Marchegiano P, Bargellini A, Cencetti S, Frezza G, Miselli M, Borella P. Effectiveness of different methods to control legionella in the water supply: Ten-year experience in an Italian university hospital. The Journal of Hospital Infection. 2011;**77**:47-51

[46] Goutziana G, Karanika VAM, Kavagias A, Stathakis NE, Gourgoulianis K, Kremastinou J, Hadjichristodoulou C. Legionella species colonization of water distribution systems, pools and air conditioning systems in cruise ships and ferries. BMC Public Health. 2008; **8**:390-397

[47] Storey MV, Ashbolt NJ, Stenström TA. Biofilms, thermophilic amoebae and *Legionella pneumophila*—A quantitative risk assessment for distributed water. Water Science and Technology. 2004;**50**(1):77-82

[48] Scheikl U, Sommer R, Kirschner A, Rameder A, Schrammel B, Zweimüller I, et al. Free-living amoebae (FLA) co-occurring with legionellae in industrial waters. European Journal of Protistology. 2014;**50**(4):422-429

[49] https://www.wur.nl/en/show/Elucidation-of-relationships-between-protozoa-Legionella-pneumophila-and-biofilm-concentrations-in-tap-water-installations.htm [Accessed: 2017-2112-17]

[50] Wang H, Edwards M, Falkinham JO, Pruden A. Molecular survey of the occurrence of Legionella spp. Mycobacterium spp. *Pseudomonas aeruginosa* and amoeba hosts in two chloraminated drinking water distribution systems. Applied and Environmental Microbiology. 2012;**78**(17):6285-6294

[51] Rozentale B, Bormane B, Perevoscikovs J, Lucenko I, Increase BA. of cases of legionelosis in Latvia. Euro Surveillance. 2011;**16**(45):1-3

[52] Fields BS, Benson RF, Besser RE. Legionella and Legionnaires' Disease: 25 years of investigation. Clinical Microbiology Reviews. 2002;**15**(3):506-526

[53] Campese C, Bitar D, Jarraud S, Maine C, Forey F, Etienne J, Desenclos JC, Saura C, Che D. Progress in the surveillance and control of Legionella infection in France, 1998–2008. International Journal of Infectious Diseases. 2011;**15**(1):e30-e37

[54] Ricketts KD, Joseph CA. The distribution of travel associated Legionnaires' disease within selected European countries, and a comparison with tourist patterns. Epidemiology and Infection. 2005;**22**:1-7

[55] Erdoğan H, Arslan H. Colonization of *Legionella* Species in Hotel Water Systems in Turkey. Journal of Travel Medicine. 2007;**14**(6):369-373

[56] Bagh LK, Albrechtsen HJ, Arvin E, Ovesen K. Distribution of bacteria in a domestic hot water system in a Danish apartment building. Water Research. 2004;**38**(1):225-235

[57] Erdoğan H, Arslan H. Evaluation of a Legionella outbreak emerged in a recently opening hotel. Mikrobiyoloji Bülteni. 2013;**47**(2):240-249

[58] Smoljanović M, Epidemiološke osobitosti zaštite zdravlja turista. Hrvatski časopis za javno zdravstvo. 2007;3(10): http://www.hcjz.hr/index.php/hcjz/article/view/2073/2046

[59] Laganà P, Gambuzza ME, Delia S. Legionella risk assessment in cruise ships and ferries. Annals of Agricultural and Environmental Medicine. 2017;**24**(2):276-282

[60] Andersen E, Lřfsgaard BE, editors. Water Report 120 Safe, Sufficient and Good Potable Water Offshore, A guideline to design and operation of offshore potable water systems. 3rd ed. Norwegian Institute of. Public Health. 2013:59-60

[61] Faezel LM, Baumgartner LK, Peterson KL, Frank DN, Harris JK, Opportunistic PNR. pathogens in showerhead biofilms. Proceedings of the National Academy of Sciences of the United States of America. 2009;**106**(38):16393-16398

[62] Blecken GT, Zinger Y, Muthanna TM, Deletic A, Fletcher TD, Viklander M. The influence of temperature on nutrient treatment efficiency in stormeater biofilter systems. Water Science and Technology. 2007;**56**(10):83-91

[63] Boltz JP, La Motta EJ. Kinetics of particulate organic matter removal as a response to biofl occulation in aerobic biofilm reactors. Water Environment Research. 2007;**79**(7):725-735

Chapter 3

Water Pollution: Effects, Prevention, and Climatic Impact

Inyinbor Adejumoke A., Adebesin Babatunde O.,
Oluyori Abimbola P., Adelani-Akande Tabitha A.,
Dada Adewumi O. and Oreofe Toyin A.

Additional information is available at the end of the chapter

http://dx.doi.org/10.5772/intechopen.72018

Abstract

The stress on our water environment as a result of increased industrialization, which aids urbanization, is becoming very high thus reducing the availability of clean water. Polluted water is of great concern to the aquatic organism, plants, humans, and climate and indeed alters the ecosystem. The preservation of our water environment, which is embedded in sustainable development, must be well driven by all sectors. While effective wastewater treatment has the tendency of salvaging the water environment, integration of environmental policies into the actor firms core objectives coupled with continuous periodical enlightenment on the present and future consequences of environmental/water pollution will greatly assist in conserving the water environment.

Keywords: pollution, chemical pollutants, microbial pollutants, sustainable development, climate change

1. Introduction

Industrialization, in any society, is a major initiator of development and urbanization. Although the merits of industrialization are innumerable, it has been identified as a major threat to the environment as it releases various toxic chemicals, gases, solid wastes as well as microbes of various kinds into our immediate environment—land, air, and water. Of particular interest is water pollution, which has become a global challenge, developing nations being highly affected due to their drive for development [1, 2].

Pollution of our water bodies poses a great threat to humans and the aquatic ecosystem while marked population increase catalyzes climatic changes [3]. For instance, various human activities as well as the release of greenhouse gases by industries greatly contributes to global warming, planet temperature enhancement, and lowering of atmospheric air quality.

The drive for sustainable development must bring along water pollution prevention techniques. Effective wastewater treatment before their eventual discharge is one way to driving water pollution prevention. Some remediate climate change mitigation measures against water pollution can also be explored.

This chapter is aimed at critically discussing water pollution effects viz-a-viz global challenges, threat, and climatic impacts while also focusing on various possible preventive measures.

2. Water pollution

2.1. Promoting environmental sustainability

Sustainable development in any society is an access to initiate a good standard of living for the populace. It aims at providing solutions to the economic, environmental, and societal challenges without posing a threat to human and environmental future development, that is, we must consider the future as we make present decisions. Also, these include social progress and equality, environmental protection, conservation of natural resources, and stable economic growth [4].

There are numerous instances where urbanization has destroyed the environment and threatened its survival chances. Sustainable development put into consideration how we survive in the natural world protecting it from destruction and damages. One of the major challenges of urbanization is sustainability, as most developed or developing society now revitalizes a lot of natural resources daily. Most of these resources meet the needs of man but they are also limited. Sustainable development tends to balance the competing needs of the society.

In achieving this, many science bodies and institutions have seen the requisite of sustainable development and have set goals and targets to meet it. This also has pushed such institutions to have a role in measuring and monitoring the impact of these goals on the society. However, the contribution of scientist in sustainable development should not only focus on the environment. It should also take into consideration the health of the populace in ensuring that no area of life suffers [5].

While sustainable development may mean different things to different people, environmental sustainability is all encompassing. It is directly concerned with the future of humanity, and it defines how we should protect and handle the sustainability of resources, air quality, water quality, and ecosystems. It also helps to prevent the environment from impending damage from technological advancement. One way to achieving environmental sustainability is via effective wastewater treatment.

Various conventional wastewater treatment methods are available; their characteristics vary from complexities of operations through sludge generations among other things to various inadequacies. Their economic disadvantages are widely related to expensive equipment, complexities of operations and skilful manpower requirement. Many industries avoid the conventional wastewater treatment methods due to their economic disadvantages, hence discharging untreated or fairly treated wastewater into the water bodies. A simple and cheap wastewater treatment method will therefore facilitate effective wastewater treatment and protect the water environment from pollution.

2.2. Effects and challenges

Water is an essential and general need of life with an undeniable effect directly or indirectly. All industrial, environmental, and metabolic processes are water dependent. In living organisms, water plays a number of roles such as solvent, temperature buffer, metabolite, living environment, and lubricants [6]. Water, however, is said to be polluted when some of the water quality parameters have been hampered by unguided and irregularities from several anthropogenic activities, thus rendering water unfit for intended use.

Water pollution may pose serious threat to the environment as well as lives. Pollutant effects may vary depending on their types and source. For instance, while heavy metals, dyes, and some other organic pollutants have been identified as carcinogens, hormones, pharmaceuticals, and cosmetics and personal care product wastes are known as endocrine disruptive chemicals [7]. These pollutants, which enter into the water body through various channel but predominantly anthropogenic, have become a great concern to environmentalists due to various hazard they pose on the environment.

2.2.1. Heavy metal pollution

Heavy metals top the list of inorganic pollutant with wide range of negative effects on aquatic organisms, plants, and human. Heavy metals are released into the environment via different routes such as industries, mining activities, agricultural activities etc. [8]. Bioavailable metals present in the soil may be absorbed by plants resulting in serious plant metabolism dysfunctioning [9]. High heavy metal ion concentrations are also known to damage the cell membrane, affect enzyme involved in chlorophyll production, thus reducing photosynthetic rate as well as affect plant reproduction via decrease in pollen and seed viability [10].

Humans and animals can be exposed to heavy metal toxicity through the food web, direct consumption of water containing metal or via inhalation [11]. Heavy metals readily bioaccumulates in vegetables and enters into man and animal through food chain. Effects of heavy metal toxicity on human ranges from mild eye, nose and skin irritations through severe headache, stomach ache, diarrhea, hematemesis, vomiting, dizziness to organ dysfunctioning such as cirrhosis, necrosis, low blood pressure, hypertension, and gastrointestinal distress [12]. While some heavy metals also called essential elements (cobalt, copper, iron, manganese, vanadium, and zinc) are required in minute amount in the body for various bio chemical processes; others such as lead, cadmium, arsenic, and mercury are of serious threat and considered foreign in the body. Looking at specifics, human ingestion of water polluted with arsenic can cause cancer of the lungs, liver, and bladder. Kidney and lungs damage as well as bone fragility may result when cadmium containing water is ingested. Exposure to lead can severely damage the brain and kidneys. In children, lead exposure even at very low concentration may hamper learning, cause memory loss, affect attention and response functions, and generally make children aggressive [13, 14]. In pregnant women, high levels of exposure to lead may cause miscarriage, whereas in men, it can damage the organs responsible for sperm production. Mercury is unique amidst other heavy metals; it has the capacity to travel a wide range of distance, thus have been classified as a global pollutants. The chemical form of mercury in the environment is also important in analyzing their toxicity. The organic form of mercury, that is, methyl mercury (MeHg) and dimethyl mercury (DMeHg), is known to be more toxic than inorganic mercury [15, 16]. While inhaled mercury goes into the blood stream, their elimination from the body is either through the urine or faeces. Mercury has the ability to exist in the urine for about 2 months, hence their renal dysfunctioning characteristic [16].

Many physiological disorders may accompany crustaceans' exposure to metals, and instant metabolic activities' alterations. Exposure of crustaceans to heavy metals may also result in loss of appetite for food and subsequently body weight loss. Continuous exposure may reduce reproduction in adults as well as hamper the growth larvae [17].

2.2.2. *Organic pollutants*

Organic pollutants are very wide in variety with a huge range of toxicity. Among the list of organic pollutants that has been of great threat to aquatic organisms, plants, and humans are dyes, plant and animal pharmaceuticals, personal care products wastes as well as petroleum organic pollutants. A group of chemicals referred to as endocrine disruptive chemicals (EDCs) also belongs to the organic pollutants group, which are classed as emerging contaminants. EDCs are described as external agents that interfere with hormonal activities, thus affecting the normal homeostatic reproduction, development or behavior [18].

Dyes are water soluble giant chemical that is greatly used in many industries viz; textile, leather and tanning, food, paper, etc. to impact color on products. Aquatic organisms, plants, and humans are greatly affected by dyes' presence in water. They impede sunlight penetration into water bodies and reduce dissolved oxygen, thus leading to death of photosynthetic organism and other lives within the aquatic environment [1]. Humans may be exposed to dye toxicity via consumption of vegetables and fish which bioaccumulate dyes. The use of colored paper towels used in drying hands and in food preparation is another route of exposure to human [19]. Dyes are considered as carcinogenic and mutagenic, thus their removal from wastewater before disposal is ultimately important.

Human and veterinary pharmaceuticals, which are members of EDCs, are chemicals used as curative or preventive of various diseases. Veterinary pharmaceutical may also serve the purpose of increasing efficiency of food production. Pharmaceuticals are used widely and unavoidably, thus they enter into the environment through one of these routes indiscriminate disposal of hospital and household waste, landfill leaching, drainage water and sewage. Although pharmaceuticals have been found to exist in various environmental samples at the ng/l to μg/l levels, it is considered a great threat to both aquatic lives and humans [20]. The presence of pharmaceuticals in water is known to pose both acute and chronic toxicity on aquatic organisms [21, 22]. EDCs as their name implies causes abnormal endocrine activities and increase cancer risk in human. Their effects on aquatic lives may range from endocrine system disruption through the reduction in eggs and sperm cells production to feminization of female aquatics [23–25].

2.3. Salvaging the aquatic environment: demands and expectations

Water is one important part of our day to day activities and their preservation can never be overemphasized. Three quarter of the fluid in man is made of water and it forms the essential medium in which the biochemical reactions take place in human body. Water moves blood from one place to the other in the body and helps in digestion; electrically charged ions, which generate nerve signals that make the human brain possible, are also held and transported by water. Water is a good solvent and it is usually referred to as universal solvent; all the major components in cells, that is, protein, deoxyribonucleic acid (DNA) and polysaccharides are all soluble in water. Pure water is tasteless, odorless, and transparent and thus provides a habitat for aquatic plants and organisms because sunlight can reach them within the water. Though clean water is a vital commodity for the well-being of human but unfortunately, the availability of fresh water is unevenly distributed and greatly threatened where available due to problems associated with climate change, inefficient water management and pollution. Recent report says very high percentage of the world population still lacks water for human well-being and ecosystem conservation [26]. The world is faced with the dilemma of achieving balance between economic development and sustainable natural environment.

Effective wastewater treatment has been earlier identified a way of protecting the water environment with detailed discussion on effective, cheap, and accessible method of wastewater treatment [27]. Various other methods of water purification such as forward and/or reverse osmosis [28–31], precipitations [32, 33], coagulation [34–36], filtrations [37–39] modular anaerobic system [40], microbial fuel cell [41], and advanced oxidation process [42] with their attendant challenges have been reported in literature.

Various environmental policies stipulating discharge protocols exists. These policies are however not effectively executed as the industries consider them as damaging to business. The ability of industries to run with the various environmental discharge policies will sustain our natural environment.

Policy integration, that is, factoring environmental issues of concern into the core of economic development, is highly important to facilitate policy performance. The main actors in environmental issues, that is, the industries, agro firms, and populace, show very little understanding of the impact of their activities on the present and future environment. While organized periodical training concerning environmental sustainability should form part of environmental policy objectives, ensuring that these objectives are integrated into sectors plains and policies is important.

3. Microbial perspective of water pollution and remediation

Drinking water supplied to our communities is usually sourced from rivers, springs, and underground sources. Usually, some form of treatment is carried out to ensure the water is fit for drinking although some sources are somewhat free from contaminating microorganism and can be clean, for example deep well. In many developing countries, one source of water can serve many uses such as drinking, washing, swimming, bathing, etc. In the same vein, sewage can be channeled into water bodies. Sewage can be defined as used water draining out of homes and industries that contain a wide range of debris, chemicals, and microorganisms. Such water is regarded as potential health hazard to consumers or the users of other sort. A major kind of hazard is the presence of pathogenic organisms in such water. This is why water is usually treated in three phases [43, 44].

The first is to separate large matter in the water source and the second stage focuses on removing more toxic substances and other matter. The tertiary phase involves total purification of water commonly by chemical disinfection. More recently, membrane bioreactors are being used and have been found to be very efficient in removing contaminants. These are combinations of communities and high-efficiency membranes that are much more effective at removing contaminants. The role of microbes is obvious in the second stage where microorganisms actively carry out biodegradation of organic matter in the aqueous portion produced after the first stage. Biodegradation of materials, such as paper and petroleum, are by bacteria, algae, and protozoa. When water is exposed to air, soil as well as effluents, it gains saprobic microorganisms; it can also pick up pathogens such as *Cryptosporidium, Campylobacter, Salmonella, Shigella,* etc.

To monitor water for each of these pathogens may not be possible but detection of fecal contamination is an easier way of spotting contamination. In such case, when the fecal contamination is high, pathogens are believed to be present and the water is unsafe for drinking. Hence, indicator organisms are used as tools to detect fecal contamination of water. They usually inhabit the intestine of mammals and birds and can be easily identified using common laboratory procedures. To achieve water protection, it will be almost impossible to search for the pathogens themselves. Hence, certain organisms with specified criteria are used as 'indicators' of the presence of enteric pathogens in a water sample. An indicator bacterium should be applicable for analysis of all types of water; it should be found anytime enteric pathogens are present and it should thrive in the wastewater longer than the toughest enteric pathogen. In addition, such organism should not reproduce in the contaminated water because this will give exaggerated values and it should not be harmful to human begins. Other criteria are that the level of contamination should be directly proportional to the level of fecal contamination; assay procedure for the indicator organism should be highly specific and the test procedure should be easily performed.

The following are commonly used as indicators of fecal contamination in water: total coliforms, *Escherichia coli,* fecal coliforms, fecal streptococci/enterococci, coliphage, and *Clostridium perfringens*. Coliforms are members of the family Enterobacteriaceae (they include *E. coli*). They are facultative anaerobic, Gram-negative, nonsporing, rod-shaped bacteria that ferment lactose with gas formation within 48 hours at 35°C [45].

Microbial contamination of water can be detected by checking for certain organisms including heterotrophic bacteria, coliforms, and *Escherichia coli* in such samples. The work by Kora *et al.* [46] showed that heterotrophic bacteria were in abundance in the lake water sampled. They also reported that contamination by *E. coli* and coliforms were beyond the allowable limit; indiscriminate disposal of sewage into water as well as release of human excreta was implicated in the high level of indicator organisms.

Control of pathogens in water is important to prevent waterborne diseases; this can be effectively done using multiple barrier approach. Microbial treatment methods goes further than traditional municipal wastewater treatment, because it takes into consideration the removal of nutrients (e.g., nitrate and phosphate) and easily degradable organic compounds as well as the possible presence of toxic compounds and variations in pH of the wastewater. A more advanced design is required in the bioreactor to be used. Some parameters to be considered in designing a treatment system are biomass yield, nutrient addition, the supply of oxygen or other electron donor, pH control as well as kinetics, that is, biological reaction rates (biotransformation). It is important to note at high concentration many compounds of interest are toxic to bacteria being used for treatment. Also some dissolved organic and inorganic compounds may constitute inhibitors to biodegradation by the organism. Biological treatment processes may not consist of the following—lagoon treatment, activated sludge as well as fixed film bioreactors.

The lagoon treatment is long-detention time basins; but unlike activated sludge processes, they do not use solid recycle. Such treatment scheme may be in three categories: anaerobic lagoon treatment, which makes use of highly loaded lakes creating anaerobic conditions. It has been used successfully for the pretreatment of meat and poultry processing wastewater reducing the biochemical oxygen demand considerably [47].

In the case of facultative ponds, there is an aerobic surface and an anaerobic bottom. The top aerobic layer facilitates treatment of dissolved organic compounds as well as odourous compounds. This has found application in pulp and paper industries. With regards to aerated lagoons, oxygen is provided by mechanical means or diffused aeration and the solids are continuously mixed and in suspension. Biological oxygen demand (BOD_5) removal may range up to 95% [48].

Another biological treatment process of interest is the activated sludge. It is made up of an aeration basin where aeration equipment provides both oxygen and adequate mixing of wastewater to maintain a uniformly mixed liquor suspended solids (MLSS). The aeration basin is followed by a liquid–solid separation usually in a clarifier by gravity and finally the settled biomass is returned again to the activated sludge basin. Examples of aeration basin configurations are – plug-flow systems, single completely mixed basins, and basins in series. The solid retention time (SRT) is important in this treatment process. The solid retention time is the average time biomass is maintained in a biological treatment process reaction. Generally,

SRT control is temperature dependent and for a warmer climate (15–25°C), SRT should be between 4 and 9 days [49]. The clarifier is very important in the performance of activated sludge processes. It ensures that efficient clarification and thickening of mixed liquor occurs. When the readily degradable soluble biochemical oxygen demand is high in wastewaters, growth of filamentous bacteria is encouraged leading to poor sludge settlement. The use of powdered activated carbon (PAC) has been discovered to enhance the efficiency of activated sludge processes. The PAC functions by adsorbing inhibitory chemicals or adsorbing chemicals that buffer variable loads. The application between 10 and 50 mg/liter of wastewater has been proven to remove organic inhibitors of the process as well as improve nitrification since it absorbs organic compounds that can prevent this process (ammonia-nitrite/nitrate conversion) in autotrophic bacteria [50].

Apart from the aforementioned, anaerobic bioreactors are also beneficial for the industrial wastewater treatment. This is because it is cost effective and can be used for industrial wastewater with high strength. The processes in the anaerobic bioreactor lead to the production of mainly methane as well as other gases. However, there is a need to strike a balance between fermentation bacterial activity and methanogenic bacteria activity as the latter is slow growing. Advantages of anaerobic treatment include low sludge formation, production of useful product, low nutrient requirement, and more importantly less energy requirements since aeration is not necessary. In addition to the energy production, advantages of anaerobic wastewater treatment, high organic matter removal efficiency, low excess sludge production, and stable operation are characteristics of this wastewater treatment technique [51].

A most recent advancement in the biological treatment of wastewater is the use of membranes in bioreactors. In such cases, the membrane can serve three major purposes. Firstly, membranes can be used as a surface for the attachment for growth of organisms and to permit oxygen to permeate into the biofilm. An example of this is the hollow-fiber gas-permeable membranes in wastewater treatment. Such membrane is produced from microporous, hydrophobic polypropylene and allows almost 100% oxygen transfer while ensuring high biomass density within the space. The second way membranes can be used as selective barriers. Such membranes permit organic compounds in wastewater to permeate but do not transport ions into the bioreactor. Thus, it allows for the selection of biodegradable organic compounds. An example of a material used for such membrane is silicone rubber. Finally, membranes can be for biomass separation. This third category requires that the membrane be used instead of a clarifier after activated sludge treatment. When such membranes are used, the effluent produced is of high quality and less sludge. In addition, automated processing can be easily employed. The disadvantage however is the financial enormity of the investment for initial start-up as well as maintenance [50, 52, 53].

3.1. Monitoring water distribution systems

Since coliform bacteria are often detected in drinking water and often, the source of contamination is not known, it is important to put in place control measures. The water distribution systems must be considered because water quality deterioration (i.e., negative quality changes that occur from the point of distribution to the point of detection) may occur. This can be as a result of reduced maintenance of the distribution system or from insufficient treatment and may lead to undue microbial growth, which the consumer may not notice. Water distribution systems should be periodically flushed to remove sediments, deposits as well as the growth of microorganisms within the pipe. For areas where the flow rate is low and possibly of the water becoming stale is high, a secondary disinfection using monochloramine and proper maintenance should be carried out in such as to prevent nitrification. Another point is to avoid a break in the distribution system especially during construction, repairs or installations, and cross connections. The officers should also ensure that the level of treatment a water sample is given is in conformity with the quality of the source of water. Also, the sampling for laboratory analysis must also be taken into consideration and monitored thoroughly when aseptic techniques are compromised, detection of coliforms may occur.

Even though reports of water diseases have been low and less serious in most developed countries, it is still a major concern in some underdeveloped countries especially war-ravaged countries. It is however important to operate a multibarrier approach, which will ensure protection of the water source, and also certify adequate treatment and distribution of water. It is however important that every occurrence of coliforms in drinking water be properly investigated so that if the contamination is as a result of operational deficiency, this can be addressed and future occurrence is prevented, thus safe guarding the health of the public [54].

3.2. Herbal disinfection of water

Several modern methods of water purification have been well embraced in our society today. However, some rural dwellers who may not be able to afford these modern treatment methods still have water pollution as a major challenge [55]. Furthermore, the disinfection by-products which remain after treatment is another reason why herbal attempts in water treatment should be encouraged.

It is important to note that not many researchers apply their antimicrobial extract or fractions directly in water treatment. Many groups stop at establishing the antimicrobial potential of their study plant, whereas others go further to apply the extracts in water treatment. For instance, a reported work used alcoholic, aqueous, and fresh juice extracts of *Ocimum sanctum* (tulsi) and *Azadirachta indica* (neem) and applied them in vitro against salmonella, which was chosen as an indicator organism. The alcoholic extract gave the best result for well water, whereas the aqueous extract was best for lake water [56]. Similarly, inspired by the fact that tulsi, neem, and amla are used to treat microbial infection without any side effect, another researcher compared the effectiveness of these three herbs in water purification, using percentage of *E. coli* removal to measure the effectiveness of each herb. A notable observation is the fact that a mixture of 1% concentration of each herb is not as efficient as the synergistic combination of the three [57].

An indirect application of herbs in water purification is their use in the synthesis of nanoparticles, which are afterwards applied to remove contaminants from water [58]. These extracts influence the surface properties of the nanoparticles, thus dictating their unique properties Owing to the obvious advantages of natural disinfection, there is a need for more research into natural products for water purification. This will in no small way help rural dwellers to cheaply assess cheap clean water and so live a healthier lifestyle.

4. Water pollution and impact on climate

All organisms, including man need water for their survival. Water resource managers had strongly depended on wastewater treatment in ensuring that the quality of water is sustained, preserved, and maintained for optimal use. By 2025, an estimated around 5 billion people out of a total population of around 8 billion will be living in areas of water stress [59]. One of the major environmental issues affecting humanity is the increasing worldwide contamination of freshwater systems as a consequence of industrial and chemical compound materials being emptied into their pathways/runways, majorly in form of micro-pollutants. According to Schwarzenbach et al. [60], most of these pollutants are present at low concentrations, many of them however can raise significant toxicological concerns, more importantly when such compounds are present as constituents of composite blends. Numerous micro-pollutants had been identified in literatures [61–64], which are not vulnerable to current treatment and are subsequently transported to the aquatic environment. Some of these include steroid hormones, pesticides, industrial chemicals, pharmaceuticals, and many other emerging materials. This consequently endangers both the aquatic and human life. It is therefore not surprising that freshwater pollution is a strong public menace, which requires global concern. The next quotation properly situates the environmental risk humans are exposed to:

> *"It is in the interest of all the world that climatic changes are understood and that the risks of irreversible damage to natural systems, and the threats to the very survival of man, be evaluated and allayed with the greatest urgency"[65]*

The above quotation were the statements of the President of the Republic of the Maldives, His Excellency Maumoon Abdul, Gayoom, during the United Nations General Assembly held in 1987 in the United States of America, as adapted from the (World Health Organization Geneva Report) [66]. The meeting was centered on Issues of Environment and Development. Due to climate change effect, both the thermal and hydrological phases of rivers are expected to vary. Owning to these, it is necessary to briefly discuss what climate change is as climate change has the potential of imposing additional pressures in some regions of the world.

4.1. What climate change is

In other to get a good grasp of what climate change is, it may be better to first define climate. Climate is usually narrowly defined as the average weather, or broadly, as the statistical description in terms of the average and variability of relevant parameters or quantities of interest over a period of time, ranging from days to millions of years. Most often, the parameters often used are temperature, precipitation, wind, etc. Climate can therefore be generally described as a state, including statistical description, of the climate system.

Climate change, on the other hand, often referred to as global warming, is the rise in the average temperature on the Earth's surface. It is well believed that the climate change event is as a consequence of human use of fossil fuels, which consequently releases carbon dioxide (CO_2) and other greenhouse gases into the air. These gases trap heat within the earth's atmosphere and can have a variety of effects ranging from rising sea levels to severe weather events. Green et al. [67] had submitted that global climate change can include natural and anthropogenic influences on terrestrial climate and the hydrologic cycle. Most notable international scientific community had actualized the reality of climate change. Some of these include the National Aeronautics and Space Administration (*NASA*), the National Oceanic and Atmospheric Administration (*NOAA*), and the Environmental Protection Agency (*EPA*) of the United States, to mention a few. These agencies further reiterated that climate change is a menace created from human activity to affect human. **Figure 1** adapted from World Health Organization Geneva Report [66] revealed that all the major activities of global climatic changes eventually bore its consequence on human in general and on human health in particular.

However, pressures and anxiety on the Earth's climatic system are having impacts on the surface of the Earth. Apart from the rising surface temperatures, the activities of increasing and frequent flooding and droughts and the changes in our natural water ecosystem are other areas of great concern in the continuous existence of man on Earth. According to WHO Geneva Report [66], climatic situation and activities affect human well-being both directly and indirectly. The direct influence is through the physical effects of climatic extremes, whereas the indirect means include the influence on the intensities or level of pollution in the air, on the marine and freshwater systems that provide food and water, as well as the pathogens that cause infectious diseases. For the purpose of this section, we concentrate on the climatic activities with respect to water pollution. This will lead us to the next sub-section.

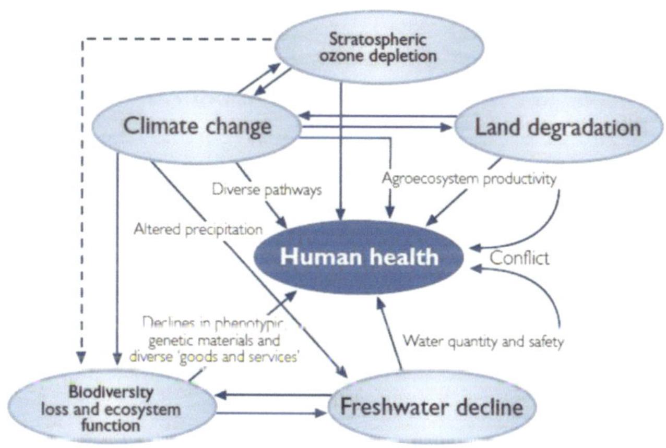

Figure 1. Inter-relationship between various kinds of environmental change. Adapted from WHO Geneva Report [66].

4.2. Probable link between water pollution and climatic impact

> *"Observational records and climate projections provide abundant evidence that freshwater resources are vulnerable and have the potential to be strongly impacted by climate change, with wide-ranging consequences for human societies and ecosystems"*

- Excerpt from the executive summary of the technical paper of the IPCC edited by Bates et al. [68]

Human activity affects weather, climate, and the environment. While some of human activities are harmless, others damage the environment. While the environment can absorb some abuse without long-term effects, much harmful human activity exceeds the environment's capacity to recover. Water pollution is one of the inevitable human-induced climate change issues that called for urgent remedial measures. Water pollution will in no small measure affect or alter the basic water quality parameters comprising the micro-pollutants, physiochemical, and biological parameters [69]. Probable and incessant changes in both rainfall and air temperature has the capacity to affect river flow thereby inducing chemical reaction kinetics as well as drop in the freshwater ecological quality. Associated with such process are dilution of contaminants and water sediment loads, which when ran into lakes will alter its natural features and affect its inhabitants. This form of water pollution or through man-made toxic chemical or/and by-products addition may therefore generate some toxic and greenhouse gases, which may subsequently contribute to global warming activities or more severe environmental threats.

The greenhouse gases are the gaseous constituents of the atmosphere (both natural and anthropogenic), which can absorb and emit radiation at certain wavelengths within the spectrum of thermal infrared radiation emitted by the earth's surface, the atmosphere itself, and the clouds. The primary greenhouse gases in the Earth's atmosphere include carbon dioxide (CO_2), nitrous oxide (N_2O), water vapor (H_2O), methane (CH_4), and ozone (O_3). The other ones identified from the Kyoto Protocol include hydrofluorocarbons (*HFCs*), perfluorocarbons (*PFCs*), and sulfur hexafluoride (SF_6). Some of these environmental threats include earth's temperature enhancement (as earlier stated), lowering of atmospheric air quality, and killing of aquatic animals. Consequently, given the legacy of historic greenhouse gas emissions and the prospect of inevitable climate change, one cannot but commit significant financial and technical resources to remediating the effect through rigorous research efforts and sensitization activities [70], more importantly to water pollution and water-related issues. Thus, water resources managers are continuously and increasingly looking for information on the possible changes in hydrological regimes, which may arise in the next few decades for likeable adaptation measure plan [68].

An interesting challenge is that while incessant water pollution may bring about a change in climatic conditions through greenhouse effect and activities, the climate too will in turn take its toll effect on the water system and environment (hydrological cycles). This is because the higher temperature generated from the greenhouse effect will eventually turn some part of the snowfall into rainfall, causing an earlier snowmelt season [71]. These effects will consequently alter the timing and volume of spring flood appreciably. The rise in sea level during this time will then cause saline water intrusion into groundwater aquifers close to the coast thereby reducing the available groundwater resources. This process will in no small measure affect humans as almost about 50% of the world population depends on groundwater for their various activities [72]. Mitchelle et al. [73] using physically hydrological and water temperature modeling framework had reported an increase in the seasonality of river discharge for about 35% of the global (consisting of all continents) land coverage for the projected and modeled data for years 2071–2100 when compared with the years spanning 1971–2000. They also projected a rise in temperature—revealing a projected global change into the future. Other effects of climate change on the hydrological cycle include increasing atmospheric water vapor content, changes in soil moisture and runoff, and changing precipitation patterns [74]. Furthermore, higher temperatures of freshwater and changes in extremes, including floods and droughts can also intensify many forms of water pollution. The authors also showed that water fluctuation in quantity and quality has the potential of affecting food stability and availability—leading to a reduction in food security to vulnerable poor farmers.

4.3. Water pollution remediates measures against climate change

Almost 20% of the entire population of the Earth do not have adequate access to safe water, while about 40% suffer the effects of deplorable sanitary situations [75]. Schwarzenbach et al. [60] had submitted that the increase in surface/groundwater pollution will go a long way in affecting both the human and aquatic life systems. According to Schwarzenbach et al. [60], about 35% of the Earth's available and renewable freshwater are used for industrial, domestic, and agricultural purposes. However, if this activities are not well managed, water pollution may be inevitable with different kinds of pollutants. Some of the water pollutants according to the authors include industrial chemicals (e.g., solvents and petrochemicals), biocides (e.g., pesticides and nonagricultural biocides), natural chemicals (e.g., heavy metals and cyanotoxins), industrial products (e.g., additives and lubricants), as well as consumer products (e.g., detergents and personal care products). Industrial manufactured nanomaterials had also been found to be a major water pollutant whose effects is on the increase in affecting aquatic ecosystem [76]. According to Lapworth et al. [77], groundwater pollution mainly results from landfill leachate, infiltration of contaminated water from agricultural land, groundwater-surface water reaction, as well as seepage of sewer systems. While some of these pollutants, like heavy metals, are not degraded at all, some others disintegrated very slowly and can be transported to hundreds of kilometers away from the source. The effects of their waterborne pollution however ranges from contamination of drinking water causing drinking water quality problems to emission of greenhouse gases, resulting in climatic change challenges.

Some remediate climate change mitigation measures against water pollution can however be taken. Some of these include carbon dioxide capture and storage, planting of bio-energy crops, proper solid waste disposition, afforestation or reforestation, cropland management–both for water and reduced tillage [74] among other measures. Scientific researches involving water pollution should be geared more toward ascertaining the physical underlying molecular mechanisms and factors rather than just the usual empirical comparison approach that is commonly used. The mechanisms involved if well understood will go a long way in properly situating ways of combating the water pollution challenges. This according to Metz and Ingold [62] may be achieved by developing a structure that addresses both the problem dimension (causes and effects) and sustainability dimension (long term and cross-sectoral) of assessing best instruments that regulate water pollution. A sustainable working policy on water pollution should not only be designed and enacted but also rigorously followed, more importantly in the developing and underdeveloped countries where the menace of water pollution has not been effectively managed. Bemelmans-Videc [78] had presented a classification for effectively managing water pollution. These are (i) regulation (i.e., substance ban and authorization restriction) (ii) economic instruments (i.e., product or substance charge—in which case a charge is levied on substances that contains hazardous compounds with the aim of reducing its use), and (iii) information (e.g., disposal requirements and information campaign). These Bemelmans-Videc [78] classifications according to Metz [63] would help in differentiating source-directed measures from end-of-pipe measures in water quality regulation. While the source-directed measures help in avoiding pollution before toxic and injurious chemical materials enter into waters, the end-of-pipe measures concentrate on filtering pollution after its input into wastewater.

Furthermore, usage and disposal approaches should be in place with the aim of minimizing the addition of critical pollutants into aquatic environment. In addition, such system-specific properties and reactivities (like adsorption to solid phases and abiotic/biological transformations) should be well understood and quantified. This kind of processes will yield a significant framework for reliable coverage and evaluation of chemical compounds in complex macroscopic ecosystems [66]. Other alternative methods for removing pollutants from waters include ozonation and advanced oxidation process (AOPs), coagulation-flocculation, membrane bioreactor, *PAC*, and attached growth treatment processes. Refer to the work of [61] for comprehensive reading on the methodology and activity of each process. It can therefore be concluded that whatever form of measure taken in inhibiting both surface and groundwater will not only advance our aquatic ecosystem, but will also help man from further endangering his environment from the menace of climate change.

5. Conclusion

Since water forms a core of the existence of human and other living things, its preservation and sustainable availability cannot be overemphasized. The availability of clean water is greatly threatened by various human activities and of interest is pollution which in turn affects the ecosystem and causes various climatic changes. While various wastewater treatment methods are being explored by industries and various treatment plants, untreated wastewater is still being discharged into the water bodies by some industries. Thus, effective environmental protection policies compliance drive will be of immense benefit to the environment and by extension to human. Factoring these environmental protection policies into the goals and objectives of various actors involved in environmental deterioration will help policies performance. This will serve as a step forward in the direction of ameliorating water pollution.

Author details

Inyinbor Adejumoke A.[1]*, Adebesin Babatunde O.[1], Oluyori Abimbola P.[1], Adelani-Akande Tabitha A.[2], Dada Adewumi O.[1] and Oreofe Toyin A.[3]

*Address all correspondence to: inyinbor.adejumoke@landmarkuniversity.edu.ng

1 Department of Physical Sciences, Landmark University, Omu Aran, Nigeria

2 Department of Biological Sciences, Landmark University, Omu Aran, Nigeria

3 Department of Chemical Engineering, Landmark University, Omu Aran, Nigeria

References

[1] Inyinbor AA, Adekola FA, Olatunji GA. Liquid phase adsorption of Rhodamine B onto acid treated *Raphia hookerie* epicarp: Kinetics, isotherm and thermodynamics studies. South African Journal of Chemistry. 2016;**69**:218-226

[2] Rana RS, Singh P, Kandari V, Singh R, Dobhal R, Gupta S. A review on characterization and bioremediation of pharmaceutical industries' wastewater: An Indian perspective. Applied Water Science. 2017;**7**:1-12

[3] Palmate SS, Pandey A, Kumar D, Pandey RP, Mishra SK. Climate change impact on for est cover and vegetation in Betwa Basin, India. Applied Water Science. 2017;**7**:1-12

[4] Ilin I, Kalinina O, Iliashenko O, Levina A. Sustainable urban development as a driver of safety system development of the urban underground. Procedia Engineering. 2016;**165**: 1673-1682

[5] Frone D-F, Frone S. The importance of water security for sustainable development in the Romanian agri-food sector. Agriculture and Agricultural Science Procedia. 2015;**6**: 674-681

[6] Hanslmeier A. Water in the universe. Astrophysics and Space Science Library. 2011;**368**. DOI: 10.1007/978-90-481-9984-6_2 © Springer Science+Business Media B.V. 2011

[7] Adeogun AO, Ibor OR, Adeduntan SD, Arukwe A. Intersex and alterations in the reproductive development of cichlid, Tilapia Guineensis, from a municipal domestic water supply lake (Eleyele) in south western Nigeria. Science of the Total Environment. 2016;**541**:372-382

[8] Cao Y, Zhang S, Wang G, Li T, Xu X, Deng O, Zhang Y, Pu Y. Enhancing the soil heavy metals removal efficiency by adding HPMA and PBTCA along with plant washing agents. Journal of Hazardous Materials. 2017;**339**:33-42

[9] Lajayer BA, Ghorbanpour M, Nikabadi S. Heavy metals in contaminated environment: Destiny of secondary metabolite biosynthesis, oxidative status and phytoextraction in medicinal plants. Ecotoxicology and Environmental Safety. 2017;**145**:377-390

[10] Xun E, Zhang Y, Zhao J, Guo J. Translocation of heavy metals from soils into floral organs and rewards of *Cucurbita pepo*: Implications for plant reproductive fitness. Ecotoxicology and Environmental Safety. 2017;**145**:235-243

[11] Popa C, Petrus M. Heavy metals impact at plants using photoacoustic spectroscopy technology with tunable CO_2 laser in the quantification of gaseous molecules. Microchemical Journal. 2017;**134**:390-399

[12] Dada AO, Adekola FA, Odebunmi EO. Kinetics and equilibrium models for sorption of cu (II) onto a novel manganese nano-adsorbent. Journal of Dispersion Science and Technology. 2015;**37**(1):119-133

[13] Verma M, Schneider JS. Strain specific effects of low level lead exposure on associative learning and memory in rats. Neurotoxicology. 2017;**62**:186-191

[14] Sun B, Zhang X, Yin Y, Sun H, Ge H, Li W. Effects of sulforaphane and vitamin E on cognitive disorder and oxidative damage in lead-exposed mice hippocampus at lactation. Journal of Trace Elements in Medicine and Biology. 2017;**44**:88-92

[15] Liu Z, Wang L, Xu J, Ding S, Feng X, Xiao H. Effects of different concentrations of mercury on accumulation of mercury by five plant species. Ecological Engineering. 2017;**106**:273-278

[16] Li P, Du B, Chan MC, Feng X. Human inorganic mercury exposure, renal effects and possible pathways in Wanshan mercury mining area, China. Environmental Research. 2015;**140**:198-204

[17] Zhang C, Yua K, Li F, Xiang J. Acute toxic effects of zinc and mercury on survival, standard metabolism, and metal accumulation in juvenile ridgetail white prawn, Exopalaemon Carinicauda. Ecotoxicology and Environmental Safety. 2017;**145**:549-556

[18] Jung C, Son A, Her N, Zoh K, Cho J, Yoon Y. Removal of endocrine disrupting compounds, pharmaceuticals, and personal care products in water using carbon nanotubes: A review. Journal of Industrial and Engineering Chemistry. 2015;**27**:1-11

[19] Oplatowska M, Donnelly RF, Majithiya RJ, Kennedy DG, Elliott CT. The potential for human exposure, direct and indirect, to the suspected carcinogenic triphenylmethane dye brilliant green from green paper towels. Food and Chemical Toxicology. 2011;**49**: 1870-1876

[20] Archer E, Petrie B, Kasprzyk-Hordern B, Wolfaardt GM. The fate of pharmaceuticals and personal care products (PPCPs), endocrine disrupting contaminants (EDCs), metabolites and illicit drugs in a WWTW and environmental waters. Chemosphere. 2017;**174**:437-446

[21] Kar S, Roy K. First report on interspecies quantitative correlation of ecotoxicity of pharmaceuticals. Chemosphere. 2010;**81**:738-747

[22] Aguirre-Martínez GV, Okello C, Salamanca MJ, Garrido C, Del Valls TA, Martín-Díaz ML. Is the step-wise tiered approach for ERA of pharmaceuticals useful for the assessment of cancer therapeutic drugs present in marine environment? Environmental Research. 2016;**144**:43-59

[23] Akanyeti I, Kraft A, Ferrari M. Hybrid polystyrene nanoparticle-ultrafiltration system for hormone removal from water. Journal of Water Process Engineering. 2017;**17**:102-109

[24] Ng CK, Bope CD, Nalaparaju A, Cheng Y, Lu L, Bin Cao WR. Concentrating synthetic estrogen 17a-ethinylestradiol using microporous polyethersulfone hollow fiber membranes: Experimental exploration and molecular simulation. Chemical Engineering Journal. 2017;**314**:80-87

[25] Saha B, Karounou E, Streat M. Removal of 17b-oestradiol and 17a-ethinyl oestradiol from water by activated carbons and hypercrosslinked polymeric phases. Reactive & Functional Polymers. 2010;**70**:531-544

[26] Costa D, Burlandoc P, Priadi C. The importance of integrated solutions to flooding and water quality problems in the tropical megacity of Jakarta. Sustainable Cities and Society. 2016;**20**:199-209

[27] Inyinbor AA, Oluyori AP, Adelani-Akande TA. Biomass valorization: Agricultural waste in environmental protection, phytomedicine and biofuel production. In: Biomass Volume Estimation and Valorization for Energy; http://dx.doi.org/10.5772/66102. 2017

[28] Zhang B, Song X, Nghiem LD, Li G, Luo W. Osmotic membrane bioreactors for wastewater reuse: Performance comparison between cellulose triacetate and polyamide thin film composite membranes. Journal of Membrane Science. 2017;**539**:383-391

[29] Bunani S, Yörükoğlu E, Yüksel Ü, Kabay N, Yüksel M, Sert G. Application of reverse osmosis for the reuse of secondary treated urban wastewater in agricultural irrigation. Desalination. 2015;**364**:68-74

[30] Zou S, He Z. Enhancing wastewater reuse by forward osmosis with self-diluted commercial fertilizer as draw solutes. Water Research. 2016;**99**:235-243

[31] Roy D, Rahni M, Pierre P, Yargeau V. Forward osmosis for the concentration and reuse of process saline wastewater. Chemical Engineering Journal. 2016;**287**:277-284

[32] Mella B, Glanert AC, Gutterres M. Removal of chromium from tanning wastewater and its reuse. Process Safety and Environmental Protection. 2015;**95**:195-201

[33] Prazeres AR, Rivas J, Almeida MA, Patainta M, Dores J, Carvalho F. Agricultural reuse of cheese whey wastewater treated by NaOH precipitation for tomato production under

several saline conditions and sludge management. Agricultural Water Management. 2016;**167**:62-74

[34] Nair AT, Ahammed MM. The reuse of water treatment sludge as a coagulant for post treatment of UASB reactor treating urban water. Journal of Cleaner Production. 2015;**96**: 272-281

[35] Da Silva LF, Barbos AD, de Paula HM, Ramuldo LL, Andrade LS. Treatment of paint manufacturing wastewater by coagulation/electrochemical methods: Proposal for disposal and/or reuse of treated water. Water Research. 2016;**101**:467-475

[36] Kausley SB, Malhotra CP, Pandit AB. Treatment and reuse of shale gas wastewater: Electrocoagulation system for enhanced removal of organic contamination and scale causing divalent cations. Journal of Water Process Engineering. 2017;**16**:149-162

[37] Rizzo L, Fiorentino A, Grassi M, Attanasio D, Guida M. Advanced treatment of urban wastewater by sand filtration and graphene adsorption for wastewater reuse: Effect on a mixture of pharmaceuticals and toxicity. Journal of Environmental Chemical Engineering. 2015;**3**:122-128

[38] Buscio V, Marín MJ, Crespi M, Gutiérrez-Bouzàn C. Reuse of textile wastewater after homogenization-decantation coupled to PVDF ultrafiltration membranes. Chemical Engineering Journal. 2015;**265**:122-128

[39] Benito A, Garcia G, Gonzalez-Olmos F. Fouling reduction by UV-based pretreatment in hollow fiber ultrafiltration membranes for urban wastewater reuse. Journal of Membrane Science. 2017;**536**:141-147

[40] Pang H, Wu P, Li L, Yu Z, Zhang Z. Effective biodegradation of organic matter and biogas reuse in a novel integrated modular anaerobic system for rural wastewater treatment: A pilot case study. Chemical Engineering & Processing: Process Intensification. 2017;**119**:131-139

[41] Abourached C, English MJ, Liu H. Wastewater treatment by microbial fuel cell (MFC) prior irrigation water reuse. Journal of Cleaner Production. 2016;**137**:144-149

[42] Biliǹska L, Gmurek M, Ledakowicz S. Textile wastewater treatment by AOPs for brine reuse. Process Safety and Environmental Protection. 2017;**109**:420-428

[43] Cowan MK. Microbiology: A Systems Approach. New York: McGraw-Hill Education; 2015

[44] Lévesque D, Cattaneoa A, Deschamps G, Hudonc C. In the eye of the beholder: Assessing the water quality of shoreline parks around the island of Montreal through citizen science. Science of the Total Environment. 2017;**579**:978-988

[45] Gonzales-Gustavson E, Cárdenas-Youngs Y, Calvo M, Marques da Silva MF, Hundesa A, Amorós I, Moreno Y, Moreno-Mesonero L, Rosell R, Ganges L, Araujo R, Girones R. Characterization of the efficiency and uncertainty of skimmed milk flocculation for the

simultaneous concentration and quantification of water-borne viruses, bacteria and protozoa. Journal of Microbiological Methods. 2017;**134**:46-53

[46] Kora AJ, Rastogi L, Kumar SJ, Jagatap BN. Physico-chemical and bacteriological screening of Hussain Sagar lake: An urban wetland. Water Science. 2017;**31**:24-33

[47] Liu YY, Haynes RJ. Origin, nature, and treatment of effluents from dairy and meat processing factories and the effects of their irrigation on the quality of agricultural soils. Critical Reviews in Environmental Science and Technology. 2011;**41**:1531-1599

[48] United State Environmental Protection Agency. Wastewater technology fact sheet, Facultative lagoon. Available from: https://www3.epa.gov/npdes/pubs/faclagon.pdf

[49] Shahzad M, Khan SJ, Paul P. Influence of temperature on the performance of a full-scale activated sludge process operated at varying solids retention times whilst treating municipal sewage. Water. 2015;**7**:855-867

[50] Encyclopedia of Microbiology. 2nd ed. 4; 2000, Editor – In Chief, Joshua Lederberg, Elsevier, India

[51] Dvořák L, Gómez M, Dolina J, Černín A. Anaerobic membrane bioreactors-a mini review with emphasis on industrial wastewater treatment: Applications, limitations and perspectives. Desalination and Water Treatment. 2016;**57**:19062-19076

[52] NSA NSAM, Zainura Zainon Noor ZZ, MAA MAAH, Gustaf Olsson G. Application of membrane bioreactor technology in treating high strength industrial wastewater: A performance review. Desalination. 2012;**305**:1-11

[53] Sharrer MJ, Rishel K, Summerfelt ST. Evaluation of a membrane biological reactor for reclaiming water, alkalinity, salts, phosphorus, and protein contained in a high-strength aquacultural wastewater. Bioresource Technology. 2010;**101**:4322-4330

[54] Bitton G, editor. Encyclopedia of Environmental Microbiology, Vol. 5. New York: John Wiley and Sons Inc.; 2002. pp. 2967-2975

[55] Shannon MA, Bohn PW, Elimelech M, Georgiadis JG, Mariñas BJ, Mayes AM. Science and technology for water purification in the coming decades. Nature. 2007;**452**:301-310

[56] Tanushree B, Milind RG, Bipinraj NK. Disinfection of drinking water in rural area using natural herbs. International Journal of Engineering Research and Development. 2013;**5**(10):7-10

[57] Rajesh R, Wankhade RR. Role of some natural herbs in water purification. IOSR Journal of Applied Chemistry (IOSR-JAC). 2016,**9**(3).38-39

[58] Das SK, Mandal AB. Green synthesis of nanomaterials with special reference to environmental and biomedical applications. Current Science. 2015;**108**(11):1999-2002

[59] Arnell NW. Climate change and global water resources. Global Environmental Change. 1999;**9**:S31-S49

[60] Schwarzenbach Rene′ P, Beate I Escher, Kathrin Fenner, Thomas B. Hofstetter, C Annette Johnson, Urs von Gunten, Bernhard Wehrli. The Challenge of Micro-pollutants in Aquatic Systems. Science. 25-08-2006;**313**:1072-1077. DOI: 10.1126/science.1127291

[61] Luo Y, Guo W, Ngo HH, Nghiem LD, Hai FI, Zhang J, Liang S. A review on the occurrence of micro-pollutants in the aquatic environment and their fate and removal during wastewater treatment. Science of the Total Environment. 2014;**473-474**(March):619-641

[62] Metz F, Ingold K. Sustainable wastewater management: Is it possible to regulate micropollution in the future by learning from the past? A policy analysis. Sustainability. 2014, 1992-2012;**6**. DOI: 10.3390/su6041992

[63] Metz F. Addressing Micropollution by Linking Problem Characteristics to Policy Instruments; Eawag ESS Working Paper Series. Dübendorf, Switzerland: Eawag; 2013, 2013 Volume 4

[64] Hering JG, Ingold KM. Water resources management: What should be integrated? Science. 2012;**336**:1234-1235

[65] His Excellency Maumoon Abdul, Gayoom, President of the Republic of the Maldives (1987): Address to the United Nations General Assembly on the issues of environment and development. New York, USA, 19-10-1987

[66] World Health Organization, Geneva (2003): Climate change and human health: Risks and responses/editors: McMichael AJ, Campbell-Lendrum DH, Corvalán CF, Ebi KL, Githeko AK, Scheraga JD, Woodward A, ISBN 92 4 156248 X

[67] Green TR, Taniguchi M, Kooi H, Gurdak JJ, Allen DM, Hiscock KM, Treidel H, Aureli A. Beneath the surface of global change: Impacts of climate change on groundwater. Journal of Hydrology. 2011;**405**:532-560. DOI: 10.1016/j.jhydrol.2011.05.002

[68] Bates BC, Kundzewicz ZW, Wu S, Palutikof JP, editors. Climate Change and Water. Technical Paper of the Intergovernmental Panel on Climate Change, IPCC Secretariat, Geneva, 210 pp. © 2008, Intergovernmental Panel on Climate Change, ISBN: 978-92-9169-123-4; 2008

[69] Delpla I, Jung A-V, Baures E, Clement M, Thomas O. Impacts of climate change on surface water quality in relation to drinking water production. Environment International. 2009;**35**:1225-1233

[70] Prudhomme C, Wilby RL, Crooks S, Kay AL, Reynard NS. Scenario-neutral approach to climate change impact studies: Application to flood risk. Journal of Hydrology. 2010;**390**:198-209

[71] Arnell NW, Charlton MB, Lowe JA. The effect of climate policy on the impacts of climate change on river flows in the UK. Journal of Hydrology. 2014;**510**:424-435 http://dx.doi.org/10.1016/j.jhydrol.2013.12.046

[72] Taikan O, Shinjiro K. Global hydrological cycles and world water resources. Science. 25 August 2006;**313**:1068-1072. DOI: 10.1126/science.1128845

[73] World Water Day, www.worldwaterday.org/wwday/1998/

[74] MTH van Vliet, WHP Franssen, JR Yearsley, F Ludwig, I Haddeland, DP Lettenmaier, P Kabat. Global river discharge and water temperature under climate change. Global Environmental Change. 2013;**23**:450-464

[75] United Nations Educational, Scientific, and Cultural Organization (2003): World Water Assessment Programme, Water for People, Water for Life -the United Nations World Water Development Report. Barcelona: Berghahn Books; 2003

[76] Wiesner MR, Lowry GV, Alvarez P, Dionysiou D, Biswas P. Environmental Science & Technology. 2006;**40**:4336

[77] Lapworth D, Baran N, Stuart M, Ward R. Emerging organic contaminants in groundwater: A review of sources, fate and occurrence. Environmental Pollution. 2012;**163**:287-303

[78] Bemelmans-Videc M-L, Rist RC, Vedung E. Carrots, Sticks & Sermons: Policy Instruments and their Evaluation. Piscataway, NJ, USA: Transaction Publishers; 1998

Chapter 4

A Temporal Analysis of Water Quality Variability at the Seattle Aquarium in Elliott Bay, Puget Sound, WA

Amy Y. Olsen, Angela Smith and Shawn Larson

Additional information is available at the end of the chapter

http://dx.doi.org/10.5772/intechopen.71353

Abstract

The Seattle Aquarium is centrally located on Elliott Bay in Puget Sound, built on a pier along the central waterfront in Seattle, WA, USA. The Seattle Aquarium Water Quality Laboratory regularly measures water quality metrics on incoming saltwater pumped directly from Elliott Bay for use in the animal exhibits. This study provides a descriptive temporal analysis of variability in the incoming saltwater conducted from 2007 through 2016. Parameters measured on a weekly basis include ammonia (NH_3), nitrite (NO_2), pH and fecal coliform bacteria. Ammonia mean throughout the dataset was 0.02 mg/l (SE ± 0.0005), with clear seasonal trends of higher ammonia levels during the summer months (May, June and July) annually. Nitrite mean was 0.01 mg/l (SE ± 0.002), with clear seasonal trends of this nutrient with bi-annual peaks in spring and fall (May and September). Saltwater pH mean was 7.81 (SE ± 0.004), trending lower in winter and spring and higher in summer and fall. Fecal coliform bacteria mean over the 10-year period was 20 colony-forming units (CFU) per 100 ml of water. Overall, Elliott Bay water quality remained relatively stable from 2007 to 2016, and if remains unchanged, will continue to be a reliable source of saltwater with known water quality parameters for use in animal exhibits in the Seattle Aquarium.

Keywords: water quality, Puget Sound, Elliott Bay, Seattle Aquarium, ammonia, nitrite, pH, fecal coliform bacteria

1. Introduction

Water quality is fundamental for the health of aquatic organisms. Many characteristics can be used to define water quality and can be relatively easily quantified by biological, physical, chemical, and esthetic indicators. Clean, uncontaminated water is preferable if not required for sustaining all trophic levels from vegetation to apex predators, including human populations.

A reduction in relative water quality can cause cascading negative effects throughout the food web, thus disrupting stable ecosystem functioning. Urbanization and changes in land use can alter water quality by increasing runoff from point (stormwater overflow valves, drainage ditches and pipes) and non-point sources (those flowing over land). Uncontrolled or captured freshwater runoff not only increases the freshwater influx into the marine ecosystem, but can also pick up pollutants and contaminants along the way moving them into nearshore ecosystems [1]. Loss of critical nearshore habitats, such as wetlands that act as natural buffers and water filters, negatively impact larval fish survival, resulting in higher mortalities, smaller cohorts and ultimately less food for higher predators [2].

Seattle and the surrounding communities have been listed as one of the fastest growing cities in recent years. The population has grown tremendously since first settlements in the 1800s, and the landscape has changed dramatically from a mostly coniferous forest to an urban city. The 2016 population estimate was nearly 4 million in the greater Puget Sound region and has grown 2.2% annually since 2015 [3].

The Puget Sound is an estuarine inland sea with a shoreline of nearly 4000 km located in western Washington [4]. It is fed by 14 major rivers and is surrounded by the Cascade Mountains to the east and the Olympic Mountains to the west. Elliott Bay is located in central Puget Sound and is just west of the City of Seattle with its eastern border along Seattle's waterfront. The main freshwater source into Elliott Bay is the Duwamish River. This river enters at the north side of Harbor Island, a heavily industrial area, and the general current flows northward along the coastline of the City of Seattle's waterfront lowering the salinity to estuarine conditions [5]. Tidal changes vary between 3 and 4 m, and occur approximately every 12.4 hours. Each tidal change brings in or removes about 8 km^3 of water and causes turbulent mixing within the water column [6]. Coastal development has increased in the region, with more than 30% of the shoreline reinforced with artificial bulkheads, seawalls and other structures [7]. Historically, fishing and shellfish harvesting have been important to the quality of life for people living in Puget Sound. Healthy fish and shellfish populations rely on relatively clean water and intact, natural nearshore marine ecosystems. Thus, the increased urbanization of nearshore marine areas that has occurred in the Puget Sound region is thought to have negatively impacted fish and shellfish resources [8].

The Seattle Aquarium is centrally located on the east side of the Puget Sound, directly on Elliot Bay (**Figure 1**). It is owned by the City of Seattle Department of Parks and Recreation and managed by the non-profit Seattle Aquarium Society. It is the ninth largest aquarium in the United States by attendance and in 2016, welcomed more than 850,000 visitors. The aquarium is composed of two buildings located on Piers 59 and 60, along downtown Seattle's waterfront. The Aquarium is responsible for the care of over 30,000 animals (representing more than 350 species) in public exhibits and behind-the-scenes animal holding spaces. The exhibits are divided into four main sections: warm water tropical fish and invertebrates, local Pacific Northwest cold water fish and invertebrates, coastal birds, and marine mammals.

The Aquarium utilizes Puget Sound water for use in all of the saltwater exhibits. Incoming raw Puget Sound saltwater goes through multiple stages of processing before use, dependent on

Figure 1. Location of study area, the Seattle Aquarium within Elliott Bay, Puget Sound, Washington.

exhibit (**Figure 2**). The warm water exhibits use incoming water that is sand filtered, mechanical cartridge filtered (Lifeguard Aquatics, Cerritos, CA), treated with ozone to remove organics (DEL Industries Infinity, San Luis Obispo, CA), filtered through activated carbon, UV sterilized (Emperor Aquatics, Inc., Pottstown, PA) and then finally heated to reach ambient tropical water temperature. The cold water or temperate exhibits use water that is sand filtered and depending on the exhibit, UV sterilized. Three bird habitats and two species specific marine mammal exhibits (northern fur seals, *Callorhinus ursinus*, and northern sea otters, *Enhydra lutris kenyoni*) use only sand filtered seawater. The harbor seal, *Phoca vitulina richardsi*, exhibit uses

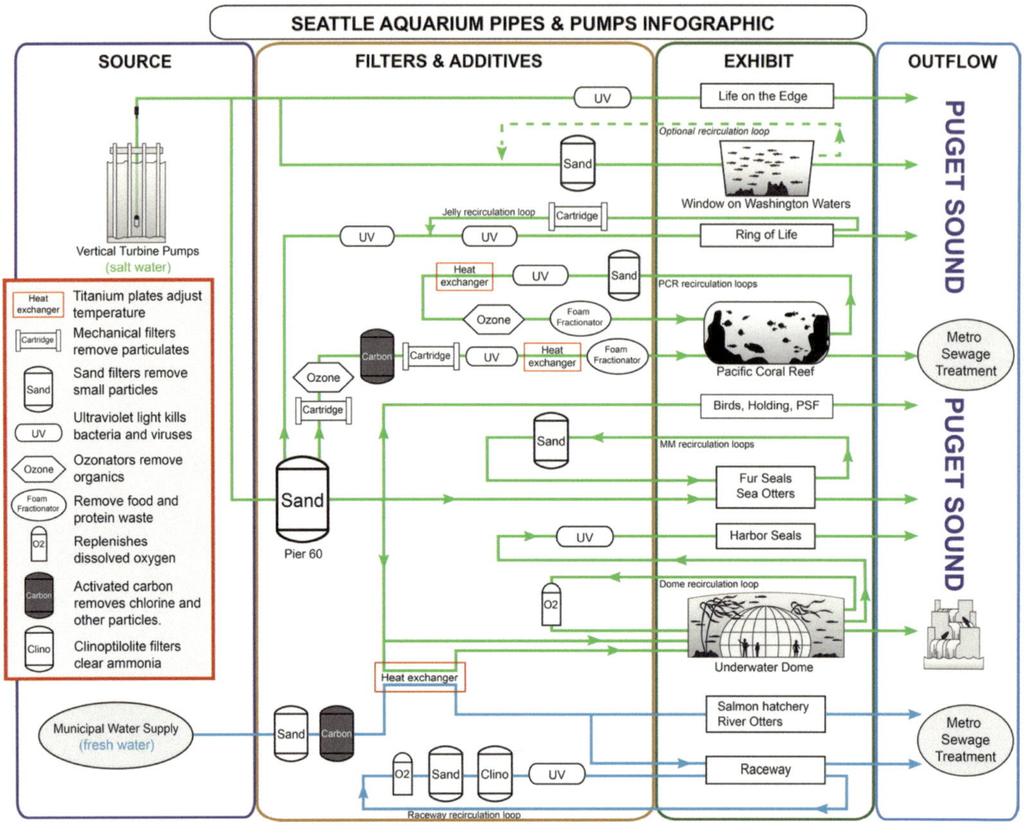

Figure 2. Pipes of pumps of the Seattle Aquarium. Incoming Puget Sound water is pumped though turbines (left) and eventually outflows to either King County metro sewage treatment or back to Puget Sound. (Source: Seattle Aquarium Engineering).

re-circulated water from the underwater dome, a temperate exhibit holding local fish, which uses Puget Sound water that is sand filtered and UV sterilized. After use in aquarium exhibits, most saltwater outflows back into Puget Sound untreated, with the exception of the water that was used in the tropical pacific coral reef exhibit which is pumped to King County Metro sewage for treatment.

The Seattle Aquarium Water Quality Laboratory has been measuring Elliot Bay water quality parameters since it opened in 1977. The principal objective for operating the laboratory is to consistently produce analytical data that accurately represents the chemical composition of the water samples. Since the aquarium is located in a highly populated and altered urban environment, it is critical to monitor the incoming water quality for use in exhibits. The results of these analyses are sent to aquarium biologists and aquarists who use this information to determine if any changes are needed (increase or decrease water flows, tank maintenance schedules, etc.) to protect and maintain aquatic animal health. Poor or inadequate water quality can weaken or sicken animals and lower disease resistance. Disease outbreaks typically occur following some form of stress to the aquatic animals, such as but not limited to changes in diet, food, tank mates, water chemistry, exposure to toxics or other environmental changes.

Water quality parameters regularly measured by the Aquarium Water Quality Laboratory and discussed here include ammonia (NH_3), nitrite (NO_2), potential of hydrogen (pH) and fecal coliform bacteria levels. Nutrients, such as NH_3 and NO_2, are key indicators of water quality. Aquatic plants require nutrients to grow and reproduce, but excessively high levels could pollute the water and cause eutrophication. This process can cause hypoxia, or very low oxygen levels, due to excessive decaying plant growth by bacteria. Potential of hydrogen (pH) is a measure of how acidic a solution is, represented by a scale from 0 to 14. Solutions with pH less than 7 are considered acidic and have a greater concentration of hydrogen ions, while those with pH higher than 7 are basic or alkaline with lower concentration of hydrogen ions. Most aquatic life have adapted to live between pH levels of 5 and 9. Fecal indicator bacteria (specifically coliform bacteria) are indicative of pollution from large amounts of human and other vertebrate sewage contamination of the water. This can result in spreading of disease and other pathogens associated with fecal contamination. Sources include leaky sewer systems, failing septic systems, pet waste, and wildlife waste. In Seattle, the Department of Health uses the Ten State Standard for beach closures. Revised in 2014, the standards are a guide for wastewater facilities to safeguard public health and protect water quality and include Indiana, Illinois, Iowa, Michigan, Minnesota, Missouri, New York, Ohio, Pennsylvania and Wisconsin [9]. King County beaches are closed to the public when coliforms exceed a geometric mean of 200 colony-forming units per 100 milliliters (CFU/100 ml) with no single sample exceeding 1000 CFU/100 ml. The Seattle Aquarium is required by the Department of Health to monitor fecal coliform bacteria in touch pools accessible to the public. In addition, the United States Department of Agriculture (USDA) requires the aquarium to monitor fecal indicator bacteria (fecal coliforms) in all marine mammal exhibits. Any sample from a marine mammal tank over 1000 CFU/100 ml is considered unacceptable and must be repeated three times every 48 hours. If the average of these samples is above 1000 CFU/100 ml then the facility must make a change to the exhibit to improve the water quality or risk losing their USDA exhibitors license permitting the facility to keep marine mammals.

Here we present the last 10 years (2007–2016) of water quality monitoring of incoming saltwater from Elliott Bay highlighting the temporal variability of these water quality parameters within Puget Sound.

2. Materials and methods

The Seattle Aquarium Water Quality Laboratory has been certified by the Washington State Department of Ecology to perform NH_3, NO_2, pH and fecal coliform analyses of water since 2014. The laboratory is inspected annually by Washington State employees and performs proficiency testing twice a year to maintain accreditation. All analytical procedures are completed according to approved standard, blank methods and include all quality assurance and quality control measures required by those methods. Quality control samples ensure that the tests performed adhere to a defined set of statistically-based criteria. These criteria are used to assess the accuracy of measurement data. Each analysis of a set of samples reported here included a check standard, blank and duplicate samples.

2.1. Water sample collection

The Seattle Aquarium has installed three intake pipes located underneath Pier 59, at 14 m below the surface of the pier (5-6 m depth in the water column, depending on tidal fluctuation). The rate of intake is 4500 gallons per minute through a 42-cm diameter pipe with 2 cm pore size screen to prevent large debris from entering. Samples reported here were raw Puget Sound saltwater collected pre-UV sterilization and filtering. Water samples were collected from a valve off the main pipe in clean plastic bottles, 2–3 times per week, from 2007 to 2016. Samples were brought back to the laboratory and are analyzed soon after collection.

2.2. Ammonia (NH_3) and nitrite (NO_2) standard method

Methods used for NH_3 and NO_2 analyses were colorimetric using a Hach (Loveland, CO) spectrophotometer to obtain absorbance values that were converted to mg/l using a conversion factor (obtained by linear regression analysis created when calibrating the spectrophotometer). Samples from 2007 through mid-2016 were analyzed on a Hach DR/5000 and the remaining samples to the end of 2016 were analyzed on a Hach DR/6000 model.

For NH_3 testing, the laboratory utilized a version of the phenol hypochlorite method outlined by Solórzano [10] and found in Standard Methods for the Examination of Water and Wastewater [11] method 4500-NH_3-F. Procedures also found in Standard Methods for the Examination of Water and Wastewater [11], were followed for NO_2 analysis (Method 4500-NO_2-B). Absorbance values for each of these tests were recorded and multiplied by a corresponding conversion factor for NH_3 and NO_2 final calculations. Duplicate deionized water blanks, a check standard (0.05 mg/l for both NH_3 and NO_2), and duplicate samples were analyzed at each session for each parameter tested. For standard preparations, an ammonia standard ampule (50 mg/l NH_3) made by Hach was utilized for years 2007 through 2013. In 2014, the laboratory switched to an ERA (Golden, CO) 1000 ± 06 mg/l NH_3 check standard which was diluted to 1.00 mg/l before use. The laboratory used a Sigma-Aldrich (St. Louis, MO) standard 1000 mg/l NO_2, also diluted to 1.0 mg/l for use, and was included for all years. Type II deionized water was used from 2007 through mid-2016 to make reagents and blanks and was produced in-house by a Barnstead Mega-Pure D1 System (Thermo Scientific, Waltham, MA). In July 2016, the laboratory switched to ultra-pure type I deionized water purchased from ChemWorld (Roswell, GA).

2.3. Potential of hydrogen (pH) method

Measurements of pH were determined using both hand-held and bench top meters. A one-point calibration was performed for the hand-held ISFET meter (Hach) using a 7.00 buffer solution prior to every analysis set. The bench top pH meter (Oakton 100 series, Vernon Hills, IL) was calibrated before each use with 4.00, 7.00, and 10.00 EMD buffer solutions. Additionally, a 6.00 + 0.05 Hach check standard buffer solution was measured to confirm the accuracy of calibration.

2.4. Fecal coliform method

The Millipore-Milliflex system (Millipore, St. Louis, MO) was used to measure fecal coliform levels. This membrane filtration system consisted of pre-made sample funnel sets for measuring

100-ml water samples. The funnels fit on top of a Millipore Plus pump platform, where 100 ml of raw saltwater samples were filtered onto a built-in 0.45 μm filter membrane. After filtration, the funnel was snapped onto a cassette filled with 2 ml of fecal coliform nutrient media (mFC dehydrated broth base with rosolic acid, EMD Millipore). The funnel was then detached, and the labeled cassette was placed in an incubator at 44.5°C for 24 hours. This method, based on Section 9222D Thermotolerant (fecal) coliform membrane filter procedure, was outlined in Standard Methods for the Examination of Water and Wastewater [11].

3. Results

Ammonia (mg/l) values from 2007 to 2016 are shown in **Figure 3** (N = 808 samples). Ammonia mean was 0.02 mg/l (SE ± 0.0004) over the 10-year period. There were clear seasonal trends as ammonia peaked during the summer months (May, June and July) annually. Typical peak levels ranged from 0.05 to 0.06 mg/l. In the winter months, levels dropped to 0.01 mg/l. Annual mean was 0.02 mg/l for years 2007 through 2012 and for 2014. Years 2013, 2015 and 2016 were 0.01 mg/l on average. The maximum level (0.11 mg/l) was recorded in June 2014. The next highest measurement, taken in July 2007, was 0.09 mg/l. Summary descriptive statistics are listed in **Table 1**, and annual statistics are listed in **Table 2**.

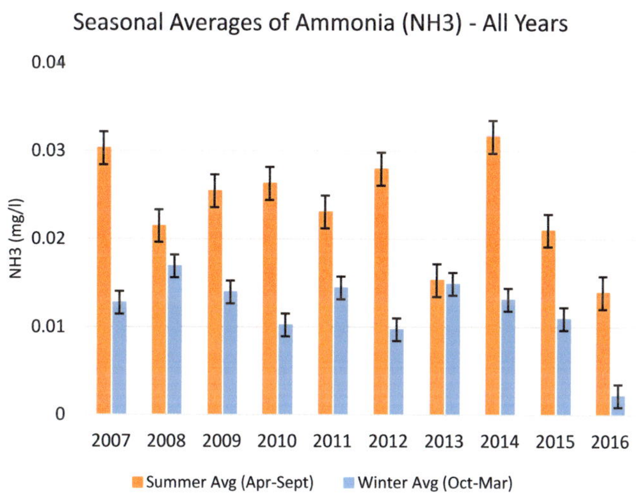

Figure 3. Ammonia (mg/l) in Puget Sound water samples from 2007 to 2016, measured in the Seattle Aquarium Water Quality Laboratory.

Nitrite (mg/l) values from 2007 to 2016 are shown in **Figure 4** (N = 811 samples). The nitrite mean was 0.00 mg/l (SE ± 0.0001) over the 10-year period. There were bi-annual, seasonal trends with peaks occurring during the late spring and fall months (April, May and September) and troughs during the winter months (December through February). The peaks were between 0.03 and 0.04 mg/l with low levels dropping to 0.00–0.01 mg/l. The annual mean was 0.00 mg/l for years 2007, 2008, 2011 through 2013, and 2016. For years 2009, 2010, 2014 and

2015, the annual mean was 0.01 mg/l. The maximum level (0.04 mg/l) for 2010 was recorded in September. The next two highest events were in September 2009 and September 2014 with 0.03 mg/l nitrite. Summary descriptive statistics are listed in **Table 1**, and annual statistics are listed in **Table 2**.

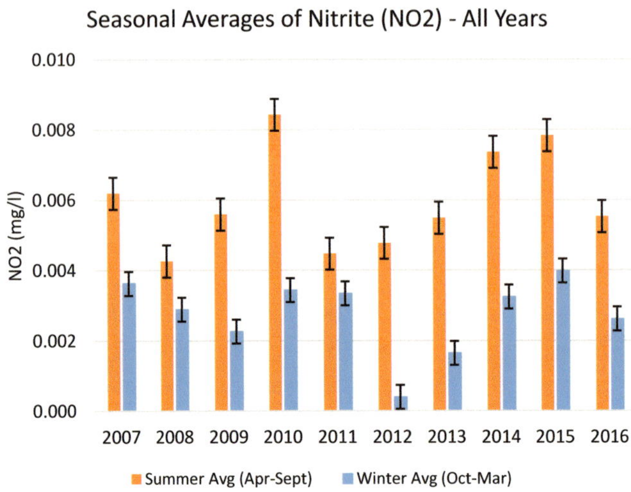

Figure 4. Nitrite (mg/l) in Puget Sound water samples from 2007 to 2016, measured in the Seattle Aquarium Water Quality Laboratory.

Water quality (2007–2016)				
	NO_2 (mg/l)	**NH_3 (mg/l)**	**pH**	**F. coliforms (CFU/100 ml)**
Mean	0.00	0.02	7.81	15
Standard error	0.00	0.00	0.00	2
Median	0.00	0.01	7.80	6
Mode	0.00	0.00	7.80	2
Standard deviation	0.00	0.01	0.11	42
Sample variance	0.00	0.00	0.01	1759
Kurtosis	13.93	5.49	0.40	91
Skewness	2.44	1.66	0.23	8
Range	0.04	0.11	0.75	584
Maximum	0.04	0.11	8.20	0
Minimum	0.00	0.00	7.45	584
Sum	3.71	14.10	6212	7616
Count	811	808	795	507

Table 1. Summary of descriptive statistics in water quality parameters, measured by the Seattle Aquarium Water Quality Laboratory, from Puget Sound water samples collected between 2007 and 2016. Parameters include nitrite (NO_2), ammonia (NH_3), pH and fecal coliforms.

Water quality by year											
		2007	2008	2009	2010	2011	2012	2013	2014	2015	2016
Nitrite (NO_2) mg/l	Mean	0.00	0.00	0.01	0.01	0.00	0.00	0.00	0.01	0.01	0.00
	SE	0.00	0.00	0.00	0.00	0.00	0.00	0.00	0.00	0.00	0.00
	Min	0.00	0.00	0.00	0.00	0.00	0.00	0.00	0.00	0.00	0.00
	Max	0.01	0.01	0.03	0.04	0.02	0.01	0.04	0.03	0.01	0.01
Ammonia (NH_3) mg/l	Mean	0.02	0.02	0.02	0.02	0.02	0.02	0.01	0.02	0.01	0.01
	SE	0.00	0.00	0.00	0.00	0.00	0.00	0.00	0.00	0.00	0.00
	Min	0.00	0.01	0.01	0.00	0.00	0.00	0.00	0.00	0.00	0.00
	Max	0.09	0.05	0.06	0.05	0.06	0.05	0.03	0.11	0.05	0.04
pH	Mean	7.85	7.79	7.84	7.87	7.80	7.82	7.79	7.79	7.81	7.80
	SE	0.02	0.01	0.01	0.01	0.01	0.01	0.01	0.01	0.01	0.01
	Min	7.45	7.60	7.70	7.70	7.50	7.60	7.60	7.50	7.60	7.58
	Max	8.20	8.06	8.10	8.20	8.00	8.10	8.00	8.10	8.00	8.20
F. coliforms (CFU/100 ml)	Mean	31	7	24	13	9	17	14	15	13	5
	SE	12	1	9	5	2	5	3	4	6	1
	Min	0	0	0	0	0	0	0	0	0	0
	Max	584	31	418	240	60	174	102	172	315	69

Parameters include nitrite, ammonia, pH and fecal coliforms.

Table 2. Descriptive statistics in water quality parameters, measured by the Seattle Aquarium Water Quality Laboratory, from Puget Sound water samples per year.

Potential of hydrogen mean was 7.81 (SE ± 0.004) from 2007 to 2016 (**Figure 5**). The pH range varied from 7.45 to 8.20 (N = 795 samples). Linear regression with negative slope of 0.00002 indicated non-significant decrease over time (R^2 = 0.03). Annual mean values are listed in **Table 2** , and monthly mean values are shown in **Figure 6**. In general, pH means fluctuated slightly year to year, with lowest values of 7.79 in 2008, 2013 and 2014. Higher values were 7.84 in 2009, 7.85 in 2007 and 7.87 in 2010. Summary descriptive statistics are listed in **Table 1**, and annual statistics are listed in **Table 2**.

Fecal coliform mean was 15 (SE ± 2) colony-forming units per 100 ml of water (CFU/100 ml) over the 10-year period of 2007 through 2016 (N = 507) (**Figure 7**). Over all 10 years in weekly monitoring, the 200 CFU/100 geometric mean maximum was exceeded three times. The single sample maximum of 1000 CFU/100 ml for Seattle beach closure was never exceeded. The lowest annual mean value was 5 (SE ± 1) CFU/100 ml in 2016, and the highest annual mean value was 31 (SE ± 812) CFU/100 ml in 2007. Minimum count throughout all years was zero, and the highest single maximum count was 584 CFU/100 ml in January 2007. Summary descriptive statistics are listed in **Table 1**, and annual statistics are listed in **Table 2**.

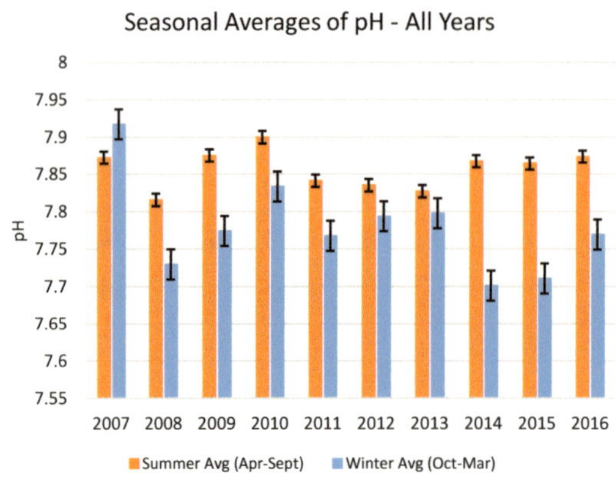

Figure 5. Puget Sound water sample pH values from 2007 to 2016, measured in the Seattle Aquarium Water Quality Laboratory.

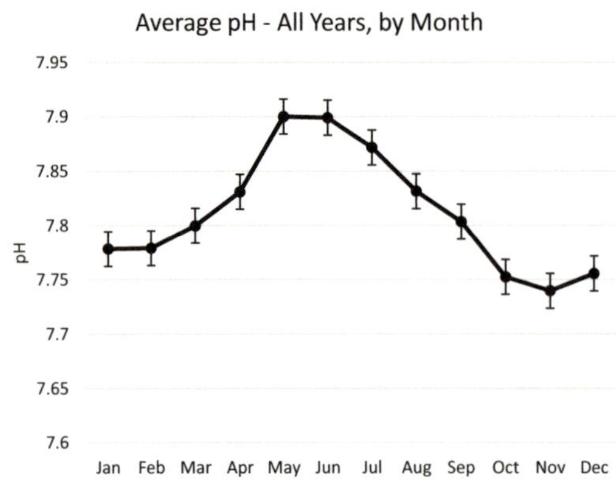

Figure 6. Monthly mean Puget Sound pH values from 2007 to 2016. Summer months increase to almost 8.0 (May and June) then drop to levels 7.75–7.8 in the fall and winter months.

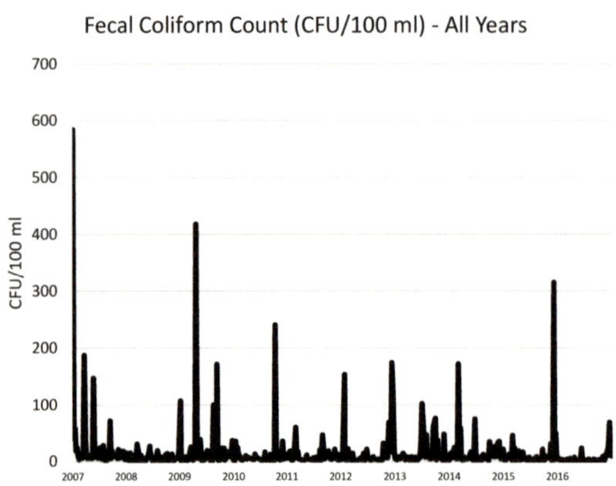

Figure 7. Fecal coliform (CFU/100 ml) levels in Puget Sound water samples from 2007 to 2016, measured in the Seattle Aquarium Water Quality Laboratory. Mean coliform count was 15 CFU/100 ml over the 10-year period.

4. Discussion

Incoming seawater parameters measured in the Seattle Aquarium Water Quality Laboratory followed expected cyclic trends from 2007 to 2016 in Elliott Bay, WA. Seasonally, there were annual phytoplankton blooms in the spring and summer months in Puget Sound, causing an increase in NO_2 followed by a subsequent increase in NH_3 as the plankton degraded. These seasonal peaks are shown in **Figures 3** and **4**. In 2011, the Puget Sound Environmental Monitoring Program (PSEMP) reported higher than average precipitation from March through May, coupled with lower temperatures, delaying the spring phytoplankton bloom [12]. The PSEMP also measured lower NH_3 levels than usual that year, with the exception of June, July and August (over the entire Puget Sound area). Our data revealed high NH_3 values in May, which decreased over the next few months. That year, NO_2 values were high only in May and September. In 2012, PSEMP reported an unusually large fall bloom in September resulting in a spike in NH_3 levels in October [13]. Our data showed a similar pattern, with a high of 0.05 mg/l in NH_3 in October of 2012. In 2013, PSEMP reported a typical spring phytoplankton bloom occurring in early April, followed by an unusually large bloom in August and September resulting in an increase in ammonia [14]. Our data does not corroborate, with no seasonal changes occurring in the incoming water at the aquarium that year. Over the course of winter 2013–2014, an anomaly called the "Blob" began. The term "Blob" refers to the mass of warm surface water that formed in the northeastern Pacific Ocean. It occurred due to a rare atmospheric circulation pattern combined with a positive Pacific Decadal Oscillation (PDO), an extended El Nino Southern Oscillation-like condition and affected Pacific climate variables (based on sea surface temperatures north of 10°N) [12, 15]. As a result, Puget Sound was influenced by warmer than normal northeast Pacific Ocean surface waters [15]. This may have stimulated an early spring plankton bloom in Puget Sound resulting in a sharp increase in NH_3 toward the end of May through June of 2014 in our dataset. The blob anomaly continued in 2015, with resulting NH_3 levels rising through late June with the exception of an October peak following a large September phytoplankton bloom [16]. Our data shows an NH_3 peak in June and again in September, followed by a decrease in October. Overall, NH_3 levels measured underneath the Seattle Aquarium tend to trend with other data measured throughout the Puget Sound.

Several factors influence the pH of Elliot Bay, including rainfall, runoff and Duwamish River input. As less dense freshwater enters the Central Sound, it tends to flow along the surface as it leaves the mouth of the Duwamish and becomes more brackish as it mixes with the denser saltwater in the bay [17]. According to Ref. [17], the Central Puget Sound receives over 8000 cfs (cubic feet per second) of freshwater inflow from the combined Puyallup, Green-Duwamish, and Cedar-Lake Washington River Basins and stormwater runoff and wastewater discharges from the Seattle-Tacoma metropolitan area. Just in Elliot Bay, discharges from the Duwamish can be as high as 2000–4000 cfs, and in late January 2011, the discharge surpassed 10,000 cfs [18].

The pH results indicate a cyclical pattern based on season from 2007 to 2016, trending toward 7.6–7.8 in the winter–spring months and 7.8–8.0 in the summer-fall months. Precipitation (rainfall and snow) in this region is higher in the winter months followed by increased freshwater release as the snow pack melts during the spring months when temperatures begin to increase. There was a non-significant downward trend noted over the monitoring period; however, it would be difficult to link this slight decrease directly to ocean acidification. The instrumentation to better monitor minute pH changes has just become available in recent years. The instrumentation utilized by the Seattle Aquarium Water Quality Laboratory to measure pH produced one or two significant figures, and the use of a more sensitive machine may better track these trends.

Incoming freshwater into Puget Sound commonly occurs through a network of rivers, streams, creeks, sewage outflow and stormwater runoff through ditches and pipes (point source pollution sites). King County utilizes combined sewer systems designed to collect rainwater runoff, domestic sewage and industrial wastewater. The water is conditioned in water treatment plants and then discharged into the Puget Sound [19–23]. There are 84 outflow locations in the City of Seattle, and four are within 2000 ft of the aquarium (No. 069, 070, 071 and 072). During high rain events, the amount of water exceeds capacity for these systems to process in the treatment plants, and when this occurs, relief point pipes release untreated sewage and stormwater into Elliott Bay, called a combined sewer overflow (CSO). These events usually increase the fecal contamination of the nearshore saltwater. Annual mean Puget Sound fecal coliform counts by the Seattle Aquarium follow a similar trend to rainfall (inches) recorded by the City of Seattle in that coliform levels tend to increase as rainfall increases (**Figure 8**). However, coliforms and precipitation do not follow a similar pattern to CSO overflow volume from the nearest four outflow pipes to the aquarium. Thus, it seems that the overflow volume of outflow nearest to the Aquarium is not indicative of coliform level in the water under the Aquarium pier. Coliform values of raw incoming seawater were typically between zero and 20 CFU/100 ml and are below critical thresholds for animal health, and in 10 years of weekly

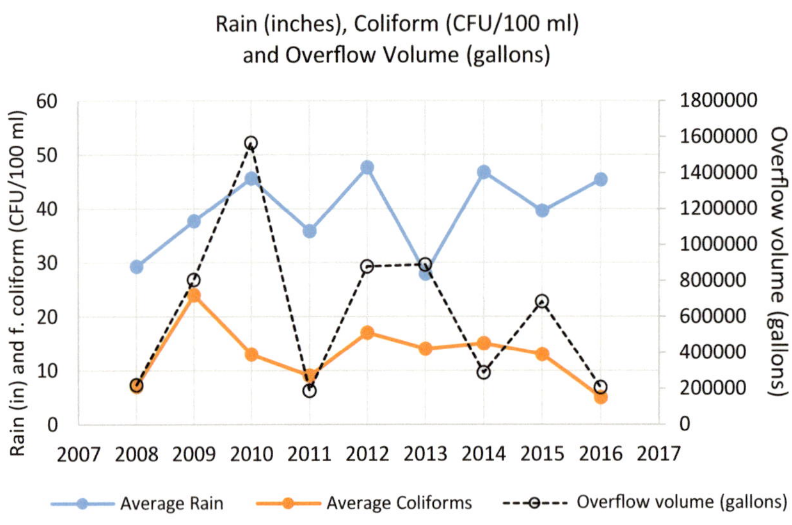

Figure 8. Mean annual rain (inches), coliform count (CFU/100 ml) and combined sewer outflow volume (gallons) from four of the nearest City of Seattle overflow locations. Coliform count and rain follow similar trends, as rain increases, coliform count also increases. Overflow volume does not seem to reflect any trends.

monitoring, fecal coliform levels never exceeded the 1000 CFU/100 ml maximum for beach closure. All touch pool exhibit water at the aquarium is UV sterilized before accessible to the public, which is also monitored weekly, to protect guests.

The stability of the water quality parameters in the incoming saltwater for the aquarium's exhibits is encouraging, given the increasing concern about climate change, ocean acidification and increasing urbanization of the Puget Sound area. The Puget Sound is home to a variety of bird, fish, invertebrate and mammal species; many categorized as vulnerable, high risk or critical to extinction risk. Flagship species determined by the Puget Sound Partnership include Pacific herring, Pacific salmon and Southern resident killer whales [24–26]. Pacific herring are considered to be the most important component of pelagic prey fish in the Puget Sound food web and a known spawning ground is just north of the Aquarium in Elliott Bay [27]. Continued stability in water quality is necessary, as it affects almost every aspect of the Puget Sound ecosystem.

Long-term longitudinal time series are essential to tease out potential seasonal or annual patterns in data. Seattle Aquarium biologists and aquarists can use this data to predict when baseline levels may shift, and use that information to fine-tune exhibit husbandry and animal care. Ongoing data collection is needed monitoring Puget Sound water quality over large time scales, as water conditions change, as urbanization increases, and as we continue to alter our marine environment.

5. Conclusions

The Seattle Aquarium is the only waterfront facility monitoring multiple water quality parameters concurrently. The King County Department of Natural Resources and Parks has a mooring buoy at the end of the Aquarium pier to measure pH, among other metrics such as temperature, salinity etc. The King County Department of Health monitors fecal coliform levels throughout the Sound. The Puget Sound Environmental Monitoring Program is a conglomerate of researchers that measure ammonia and nitrite throughout the Sound. Due to the location of the Aquarium, lab staff can quickly and easily collect samples on a day-by-day basis to provide continuous monitoring. This research provides data on water quality that is unique to Elliott Bay within the downtown Seattle area. With this data, Aquarium staff can monitor the health of the nearby marine ecosystem, and the Aquarium can serve as a watch dog to any major changes in water quality. Any values that fall outside of normal ranges can be flagged, and the proper authorities can be notified. The Aquarium lab staff work closely with the Department of Ecology, the Port of Seattle, King County and other officials, and these partnerships are essential to keep this important research going.

One shortcoming of this research is the inability to pinpoint cause. Parameters can be measured, but locating a point pollution location or determining why or how quality changed is outside of the Seattle Aquarium laboratory capabilities. Typically, lab staff collects three samples a week, and the scale of the data is limited to this time frame. Water flushes in out and out of Elliott Bay continuously, and tides that occur twice daily also provide mixing. The Aquarium intake pipes are at depth, so surface water quality is not measured, which may differ from water quality at depth.

Elliott Bay and Puget Sound are unique study areas that are estuarine, highly populated, with directed in and out flow through the straits of Juan de Fuca and Georgia. This data can contribute to understanding how a body of water like this is affected by season, rainflow and tidal fluctuation. There is limited data on similar water systems, and by contributing to this area of research, other scientists may use this data as a comparison to quantify differences between areas.

The Aquarium will continue to monitor Puget Sound water quality on a weekly basis. As incoming water is used for exhibits and animals, it is necessary to confirm that the water provided for animal health and husbandry is the best available. With a long-term dataset, trends, patterns and outliers can be identified and analyzed. This research will help with continuing to keep the Puget Sound as clean as possible and supports the Seattle Aquarium's mission of Inspiring Conservation of the Marine Environment.

Acknowledgements

We would like to thank the staff and volunteers at the Seattle Aquarium who participated in the analysis of water quality for this study. Special thanks go to their support and assistance. The Seattle Aquarium funded this research.

Author details

Amy Y. Olsen*, Angela Smith and Shawn Larson

*Address all correspondence to: a.olsen@seattleaquarium.org

Seattle Aquarium, Seattle, USA

References

[1] Weibel SR, Weidner RB, Cohen JM. Pesticides and other contaminants in rainfall and runoff. American Water Works Association. 1966;**58**:1075-1084

[2] Thom RM, Shreffler DK, Macdonald K, et al. Shoreline armoring effects on coastal ecology and biological resources in Puget Sound, Washington. Volume 7. Coastal Erosion Management Studies. 1994

[3] Puget Sound Regional Council. Regional Data Profile [Internet]. 2017. Available from: https://www.psrc.org/regional-data-profile [Accessed: 01 May 2017]

[4] Kreitler J, Papenfus M, Byrd K, et al. Interacting coastal based ecosystem services: Recreation and water quality in Puget Sound, WA. PLoS One. 2013;**8**(2). DOI: 10.1371/journal.pone.0056670

[5] Eash-Loucks W. Influence of the Duwamish River on water quality in Elliot Bay. Poster session presented at: Salish Sea Ecosystem Conference; 2014. Apr 30–May 2; Seattle, WA

[6] Rice J, Baker J, Biedenweg K, et al. Puget Sound fact book. Encyclopedia of Puget Sound. 2015;**3**(1):124. Available from: https://www.eopugetsound.org/sites/default/files/features/resources/PugetSoundFactbook_v3.1.pdf

[7] Department of Ecology. Problems with Puget Sound. No date [cited 01-06-2017]. Available from: http://www.ecy.wa.gov/puget_sound/threats.html

[8] Glasoe S, Christy A. Coastal Urbanization and Microbial Contamination of Shellfish Growing Areas. Puget Sound Action Team Publication (PSAT). Olympia, WA; 2004 Jun. p. 04-09

[9] Health Research, Inc. Recommended Standards for Wastewater Facilities. 2014. Available from: http://10statesstandards.com/wastewaterstandards.pdf

[10] Solórzano L. Determination of ammonia in natural waters by the phenol hypochlorite method. Limnology & Oceanography. 1969;**14**(5):799-801

[11] Baird RB, Eaton AW, Rice EW. Standard Methods for the Examination of Water and Wastewater. American Public Health Association, American Water Works Association, & Water Environment Federation. Washington, DC; 2012

[12] Moore S et al. Puget Sound Marine Waters Report. Puget Sound Partnership; 2011

[13] Moore S et al. Puget Sound Marine Waters Report. Puget Sound Partnership; 2012

[14] Moore S et al. Puget Sound Marine Waters Report. Puget Sound Partnership; 2013

[15] Moore S et al. Puget Sound Marine Waters Report. Puget Sound Partnership; 2014

[16] Moore S et al. Puget Sound Marine Waters Report. Puget Sound Partnership; 2015

[17] Staubitz WW, Bortleson GC, Semans SD, et al. Water-Quality Assessment of the Puget Sound Basin, Washington—Environmental Setting and its Implications for Water Quality and Aquatic Biota. Water Resources Investigations Report. USGS. Tacoma, WA; 1997

[18] Eash-Louks W. Marine Moorings in Elliott Bay: What you didn't Know you Were Missing. Power Point Presentation. King County: Marine and Sediment Assessment Group;

[19] Annual Report Combined Sewer Overflow (CSO) Reduction Program Report. Seattle Public Utilities; 2012

[20] Annual Report CSO Reduction and CMOM Programs Report. Seattle Public Utilities; 2013

[21] Annual Report. Seattle Public Utilities; 2014

[22] Wastewater Collection System: Annual Report. Seattle Public Utilities; 2015

[23] Wastewater Collection System: Annual Report. Seattle Public Utilities; 2016

[24] State of the Sound. Puget Sound Partnership. Olympia, WA; 2009

[25] State of the Sound. Puget Sound Partnership. Olympia, WA; 2012

[26] State of the Sound. Puget Sound Partnership. Olympia, WA; 2013

[27] Forage Fish Spawning Map—Washington State. 2012. Retrieved from http://wdfw.maps.arcgis.com/home/item.html?id=19b8f74e2d41470cbd80b1af8dedd6b3

Chapter 5

Challenges and Treatment of Microplastics in Water

Heloisa Westphalen and Amira Abdelrasoul

Additional information is available at the end of the chapter

http://dx.doi.org/10.5772/intechopen.71494

Abstract

Microplastics are particularly problematic and could pose big treatment. challenges. In today's world, plastic is an essential raw material. Since their invention in the 1930s, plastics have become ubiquitous in the manufacture of everyday products. Part of the problem stems from the fact that it can be difficult to pinpoint the exact source of the microplastics because of their relatively fragmented nature, small size, and a wide range of potential sources. Microplastics have become a threat to the environment, a concern reflected by sites with unusually high concentrations and a possibility of even greater concentrations in the future. Consequently, the use and subsequent release of microplastics must be drastically reduced as part of a global initiative even prior to the availability of research studies outlining the long-term risks involved. This chapter will interrogate key sources of microplastics, assess their capacity to problematize water resources in urban environments, as well as offer an analysis of its effects on the environment and some of the methodologies that can be applied to control it.

Keywords: plastic debris, environmental concern, water, personal care products, cosmetic products, microplastic, treatment

1. Introduction

Microplastic particles (MPPs) have become a global concern since they have been added to products that are used almost daily. MPPs can be found in a number of cosmetic and personal care products, including washing liquids, soaps, facial and body scrubs, toothpaste, and lotions. One of the primary issues is that the microplastics found in cosmetic and personal products are rinsed into the household drains without any precautionary recycling measures. The disposed MPPs as well as other types of plastic debris end up at the municipal wastewater treatment plants (WWTPs) because of the durability and frequent usage of synthetic polymers [1, 2]. Recent published research reports point to WWTPs as possible sources of microplastics that pollute the aquatic systems [3–5]. On the other hand, researchers could not definitively corroborate the correlation between WWTPs and microplastic pollution found in rivers. There is an ongoing debate about whether discharged effluents substantially contribute to the microplastic buildup in the , there is a lack of certainty about how such pollutants function during the wastewater treatment facilities' transport processes. Both environmental scientists and plant design engineers would benefit from an in-depth understanding of the microplastics' transport pathways and accumulation during wastewater treatment processes. A comprehensive study can help enhance and expand current treatment plant processes for coping with and eliminating this pollutant type. The exact origin of microplastics is difficult to identify because of their fragment nature, small size, and varying sources. However, the disposal of plastics into the immediate environment should be addressed in a global initiative even prior to the full environmental risk assessment. Specifically, the presence of microplastics in the environment, MMPs with overly high concentrations at certain sites, and the certainty that these concentrations will continue increasing are sufficient as a justification for a global effort. This chapter will interrogate key sources of microplastics, assess their capacity to problematize water resources in urban environments, as well as offer an analysis of its effects on the environment and some of the methodologies that can be applied to control it.

2. Origin and characterization of microplastics

According to the National Oceanic and Atmospheric Administration (NOAA) from the US Department of Commerce, plastic is the most prevalent type of marine debris found in the ocean and Great Lakes [1] (Oceanic). Its presence in aquatic environments, including beaches, ocean surface waters, deep-sea sediments, freshwater lakes, and tributaries, has been investigated during the past decades. It is a consequence of the increasing production of plastic materials and the many gaps on its proper disposal; it has become a global issue [2].

According to literature, the amount of plastic entering the ocean is increasing very fast over the years. Currently, between 0.48 and 1.27 million tons of plastic waste enters the ocean annually, and it is expected to see this number doubling in the next 10 years [3]. These materials are suitable for many industrial applications because of the presence of stable carbon-hydrogen bonds but also make it resistant to disintegrate in the environment, and hence it tends to accumulate over the time [3].

There is not much agreement between authors about the definition of microplastics, but in general, it can be referred as synthetic organic polymer particles with a size smaller than 5 mm [4]. Many published studies refer to microplastics as plastic particles or debris which are less than 5 mm length, but there is no consensus about the lower limit. Some researchers adopt 0.5 or 1 mm as a cutoff between macro- or mesoplastic and microplastic. Most of the published data refer to plastic particles ranging from 1 to 5 mm [5–9]. Ballent et al. considered three categories of microplastics: fibers, fragments, and spherical beds [2]. Common plastic polymers include polypropylene (PP), polyethylene (PE), low-density polyethylene (LDPE), and polyacrylates [5].

Many studies have been investigating the presence of plastic materials in different aquatic environments. However, some of them have been focusing on microbeads, which are primary microplastics in spherical shape. Those are mostly used in health and personal care products, and others have been studying more broad range of material classified as microplastics, including primary and secondary microplastics. Primary microplastics refer to microparticles that are manufactured in the previously mentioned scale, and secondary microplastics are the products of the degradation of larger plastic material, from mechanical or photooxidative pathways [4, 9, 10]. The lack of standardization causes difficulty in the comparison of results and the discussion of potential solutions for this environmental issue.

Primary microplastics comprehend particles used in personal care products with exfoliating purposes, such as facial cleansers and moisturizers, shampoos, cosmetics, and shaving products. Most of the microbeads in these products are composed of polyethylene (PE) and polypropylene (PP) [1, 2], and it is used for emulsion stabilization, viscosity regulation, and skin conditioning [5]. After being used, these products are washed down the drains, and microplastics are carried via wastewater to municipal wastewater treatment plants that can eventually reach the environment. Primary microplastics also include industrial abrasives or "scrubbers" used to blast clean surfaces, plastic powders used in modeling, particles used in drilling fluids for oil and gas exploration, and also raw materials used for plastic fabrication for many industrial applications [4, 5, 10, 11].

Since secondary microplastics are generated for the breakdown of larger plastic materials, there are many sources that can contribute to its presence in the environment. According to the review done by Duis and Corrs, 75–90% of the plastic found in aquatic environment originates from land-based sources and 10–25% from ocean-based sources [4]. With regard to the land-based sources, they assumed that the most important route of secondary microplastics into the environment is the loss from inappropriate managing of landfill sites and during waste collection. There are also routes related to the action of natural phenomena (such hurricanes, tsunamis, and strong sea), agricultural activities, the use of synthetic textiles, and other different human activities [4]. Based on various published studies, Duis and Corrs [4] summarized the sources of primary and secondary microplastics as shown in **Table 1**.

In 2014, Desforges and coworkers [9] documented the abundance, composition, and distribution of microplastics in the northeastern Pacific Ocean and coastal British Columbia. Considering fibers and fragments, a concentration range varying from 8 to 9200 particles per m^3 was observed. As it is shown on the map (**Figure 1**), lower concentrations were observed in offshore Pacific waters, and higher concentrations were registered nearshore with the prevalent presence of fibers. According to this study, the materials found near urban areas are likely to be from land-based sources. In specific areas, the material appeared to be composed of debris that have been trapped and concentrated by the natural ocean activity [9].

In a recent study conducted by Ballent and coworkers, the presence of microplastics in the Canadian shoreline of Lake Ontario was investigated, evaluating its abundance and distribution pattern in three depositional zones: nearshore, tributary, and beaches (**Figure 2**). The influence of microplastic pollution on benthic ecosystems and its effect in the food chain were also analyzed [2].

Primary microplastics	
Personal care products containing microplastics as exfoliants/abrasives	
Specific medical applications (e.g., dentist tooth polish)	
Drilling fluids for oil and gas exploration	
Industrial abrasives	
Preproduction plastics, production scrap, plastic degranulate: accidental losses, runoff from processing facilities	
Secondary microplastics	
General littering and dumping of plastic waste	
Losses of waste during waste collection, from landfill sites and recycling facilities	
Losses of plastic materials during natural disasters	
Plastic mulching synthetic polymer particles used to improve soil quality and as composting additive	
Abrasion/release of fibers from synthetic textiles	
Release of fibers from hygiene products	
Abrasions from car tyres	
Paints based on synthetic polymers (ship paints, other protective paints, house paint, road paint): abrasion during use and paint removal, spills, illegal dumping	
Abrasions from other plastic materials (e.g., household plastics)	
Plastic items in organic waste	
Plastic-coated or laminated paper: losses in paper recycling facilities	
Material lost or discarded from fishing vessels and aquaculture facilities	
Material lost or discarded from merchant ships (including lost cargo), recreational boats, oil and gas platforms	

Table 1. Overview of sources for primary and secondary microplastics present in the environment [4].

In this specific study, the material was counted and sorted between three categories: fibers, fragments, and spherical beads. Many factors were shown to be decisive on the distribution of microplastics. This includes the fact that this aquatic environment is surrounded by highly urbanized and industrialized areas, morphology of the shorelines, variations in topography, and so on [2]. Also, according to the author [2], the density and shape of microplastic particles may also impact distribution patterns in Lake Ontario. Even though the polymers which microplastics are made of (such as PE and PP) have lower density than the water, the material was found in the sediment, and it can be attributed the increase of the net density of microplastic particles by biofouling, adsorption of natural substances on the surface, inclusion of inorganic fillers during manufacturing, and fecal express. With regard to the shape of the material collected from Lake Ontario, it was observed that more fibers were found on the sediments and more microbeads were found on the surface water [2].

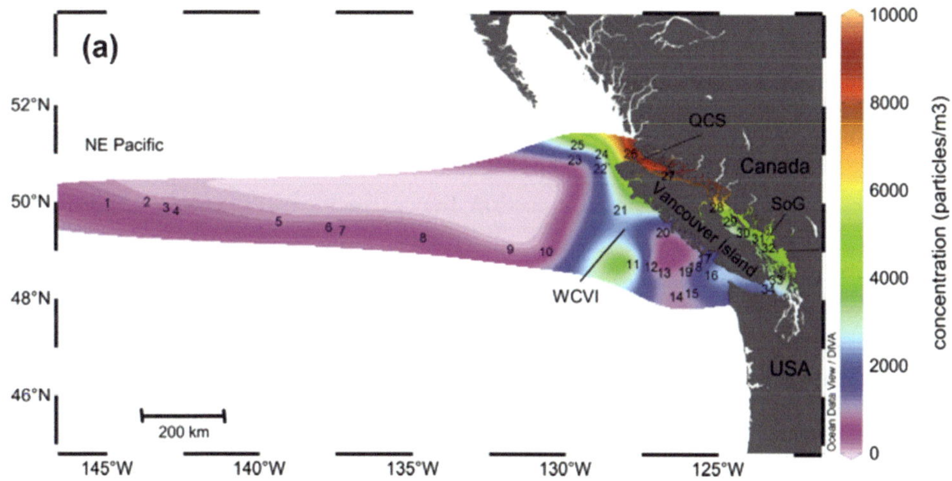

Figure 1. Total microplastic concentrations (particles/m^3; detected particles >62 μm) in subsurface waters (4.5 m) of the NE Pacific Ocean in and around coastal British Columbia, Canada [9].

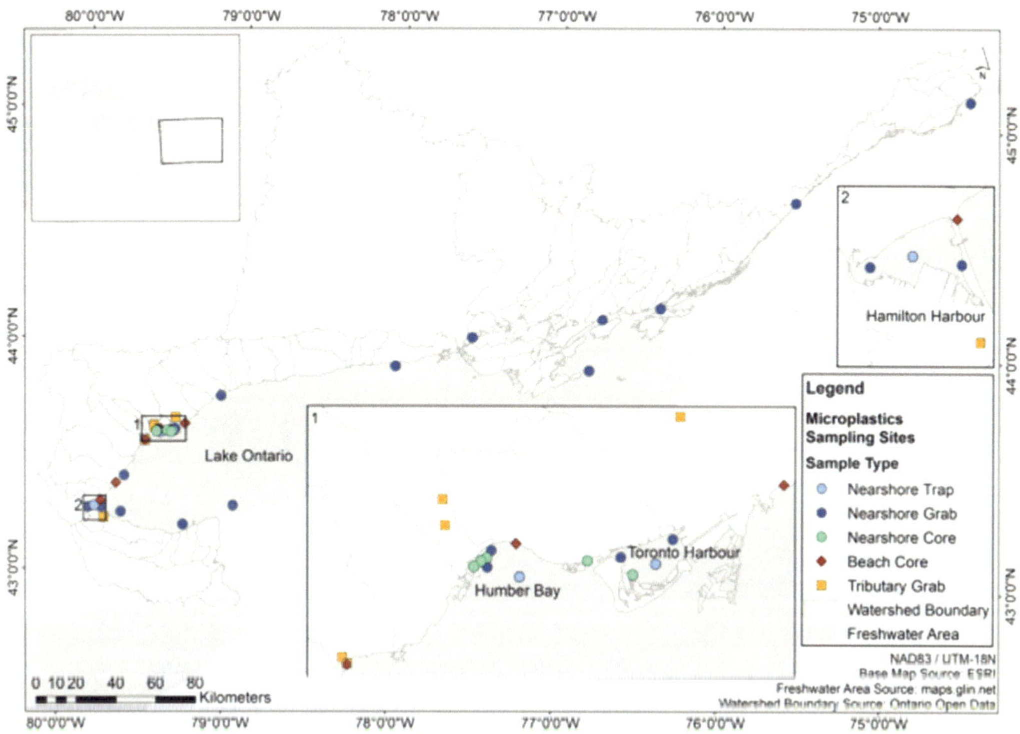

Figure 2. Sampling sites by depositional environment and instrument type for which microplastics in sediments in Lake Ontario and the St. Lawrence River were analyzed. Watershed boundaries indicate the regions that drain directly into Lake Ontario and the St. Lawrence River [2].

3. Impact of microplastics on environment

According to Anderson et al. [5], there are numerous ways through which microplastics and associated contaminants can be incorporated by the aquatic biota (**Figure 3**). This includes filter feeding, suspension feeding, inhalation at air-water surface, and consumption of prey exposed to microplastics or via direct ingestion; hence, microplastics can be assimilated into the food chain at different trophic levels [5]. Because of its particularly small size, microplastics are easily ingested by plankton and filter feeding animals which are not capable of properly selecting their food [5]. In literature, there are studies that indicate the presence of microplastics in the aquatic biota since 1960, and a few studies have been investigating the microplastic's uptake by aquatic mammals and seabirds [5].

It has been proven by many studies that microplastics are entering our food web, and as top predators, human beings are exposed to its potential harms. A study reported by Davidson and coworkers [12] investigated the presence of microplastics in Manila clams in British Columbia. It was found that every clam analyzed had at least one piece of microplastic and the concentration in individual clams ranged from 0.07 to 5.5 particles per gram. In Brazil, an investigation conducted by Miranda [13] detected significant amounts of MMP in the stomachs of two important edible species of fish. According to review published by Santillo, MMPs have been identified in different marine species from various countries around the world such as Canada, Brazil, Portugal, China, Indonesia, the USA, and Spain and important regions such as the North Pacific central gyre, North Atlantic, English Channel, North Sea, Baltic Sea, and central Mediterranean [14].

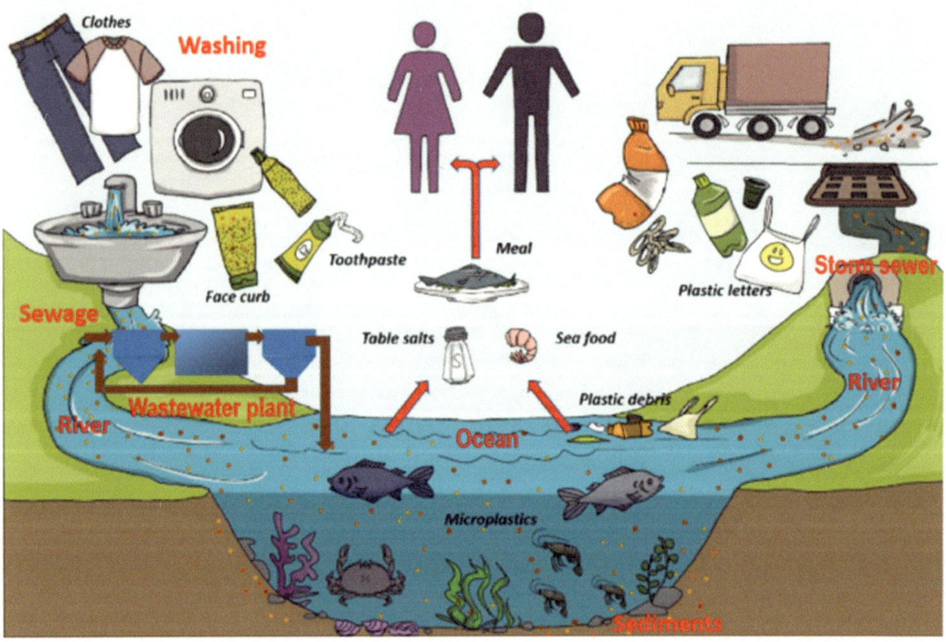

Figure 3. Microplastic pollution in aquatic environments and impacts on food chains [3].

Microplastics can be hazardous to aquatic organisms through different pathways [5, 7]. First, the ingestion of microplastics can cause physical blockage, internal abrasions, and internal and external wounds, and hence the organisms can be harmed by expending energy for egestion, can suffer from starvation and debilitation, and can result in death [3–5, 7]. Second, the organisms can be exposed to the leakage of toxic additives such as plasticizers, stabilizers, pigments, fillers, and flame retardants. These substances cannot only be toxic but also carcinogenic or endocrine active, which can impact reproductive functions of the species [5, 7, 14, 15]. Furthermore, because of its large surface-to-volume ratio and some of its inherent characteristics, microplastics are capable of absorbing various contaminants, such as heavy metals, chlorinated and aromatic compounds, and potentially persistent organic pollutants due to its hydrophobic nature [3, 5, 9, 16, 17].

In more recent studies, the ecological impact of microplastics as carriers of toxic compounds has been investigated [3, 5, 7, 9]. It has been reported that one of the many routes through which microplastics can reach the environment is through the low efficiency of water and wastewater treatment plants in retaining this material [3, 5]. Since there is a great deal of substances in these plants and microbeads can absorb and concentrate toxic hydrophobic substances, it can work as delivery systems of these substances to the aquatic environment [3].

Many research efforts have been put on trying to identify which are the main contributors for microplastic pollution. Scientists in the Ministry of the Environment and Climate Change (MOECC) have been studying the presence of microplastics in the Great Lakes [18]. They collected surface water samples in 2014 from nearshore areas in Lake Erie downstream of Detroit-Windsor, near the mouth of the Grand River, and near Fort Erie. According to the collected data, microbeads (primary microplastics) comprised approximately 14% of the material collected. Also, higher concentration of microplastics near to highly urbanized areas such as Humber Bay of Toronto was observed [18]. Greater amount of secondary microplastics were found, which includes fragments from broken down litter, shavings from cuttings/trimmings of plastic, foam from Styrofoam packaging, and fibers. It was also observed that after rainstorms occur, there was an increase of the presence of microplastics due to the drag of particles from roads and landscape [18]. They also studied the composition of the effluent form of a wastewater treatment, and it was observed that microbeads comprised 30% of the microplastics found in the effluent samples. From urban streams in the Toronto area, microbeads were found in lower amount (less than 2%), but fibers accounted for the greatest amount of microplastic particles in those samples [18].

It is very hard to determine the origin of the microplastics due to its small size, fragment nature, and unknown range of possible sources [2]. The potential origin of polyurethane particles is foam from furniture, adhesives such as construction glue products, surface coating, and sealing applications. Some particles (black, opaque with rubberlike consistency) are probably from vehicle tires, from both natural wear-down process during driving and shredding of used tires for recycling purposes. The origin of fiber material is probably from the production, washing, and natural aging of textiles (synthetic clothes and carpets). The potential source of some particles (amber-colored beads) include exchange medium for water purification and softening, as well in various medical and industrial applications [2]. Another source of microplastics in aquatic environments is the drag of this material by storm and extreme weather events.

According to the results obtained from Ballent's study in 2015, most the spherical beads (microbeads) found in sediments were from non-cosmetic source; however, it is possible that some of the fragments found in the samples originated from cosmetic products such as toothpaste and face washes. Nevertheless, the inclusion of microbeads manufactured for use in cosmetic products to the List of Toxic Substances in Schedule 1 of the Environmental Protection Act of 1999 is still important [2]. In 2013, it was found that microbeads comprised approximately 58% of the microplastics (smaller than 1mm in size) collected from the surface of the Great Lake, and this material is comparable to particles used in cosmetic products [19].

As consequence of numerous investigations on the effect of microplastics in aquatic environments, microbeads were officially declared toxic by the Canadian in June of 2016. According to the federal government, sale of shower gels, toothpaste, and facial scrubs containing microbeads will be banned in July 2018, and the use of microbeads in natural health products and nonprescription drugs will be prohibited in 2019 [20].

4. Water and wastewater treatment plants

Water and wastewater treatment plants (WWTPs) receive wastewater from households, institutions, commercial establishments, and industries and sometimes also from rainwater runoff from urban areas. WWTPs focus on the removal of large solid debris and reduce the concentration of nutrient and organic material. To complete this, a combination of physical, chemical, and biological processes is used. Some facilities also include a final treatment using disk filter or membrane bioreactor (MBR) to reduce the amount of particulate material in the effluent water [16].

Typically, a wastewater treatment plant is divided into three main steps as shown in **Figure 4**: primary, secondary, and tertiary treatments [15]. Usually, a preliminary treatment is applied to remove solid materials that can hinder the following steps of the process, which can include coarse screening, grit removal, and grinding of large objects [21]. The primary treatment uses sedimentation and skimming to remove settleable organic and inorganic solids, and besides removing income biochemical oxygen demand and oil and grease, it can reach 50–70% removal of total suspended solids [21]. The secondary treatment focuses on the removal of residual organic and suspended solids. It usually applies biological treatments such as activated sludge, trickling filters, and rotating biological contactors [21]. Some facilities apply a tertiary or advanced treatment to remove nitrogen, phosphorus, additional suspended solids, refractory organics, heavy metals, and dissolved solids [21].

Currently, the contribution of the WWTPs to the discharge of large amounts of microplastics into the environment is still a debate between the authors. Also, the lack of standardization on the methodology applied to collect and interpret data makes it more difficult to reach a common sense about the link between microplastic pollution in aquatic systems and the WWTPs.

Per published studies, there are three main contamination routes for freshwater systems: effluent discharged from WWTPs, overflow of wastewater sewers during high rain events, and runoff from sludge applied to agricultural land [5, 19]. It is estimated that approximately 80% of the microplastics in oceans originate from land-based sources and another 18% from aquaculture or fishing industries [5]. According to Ziajahromi et al., land-based sources include urban runoff and effluent discharge from WWTPs [17].

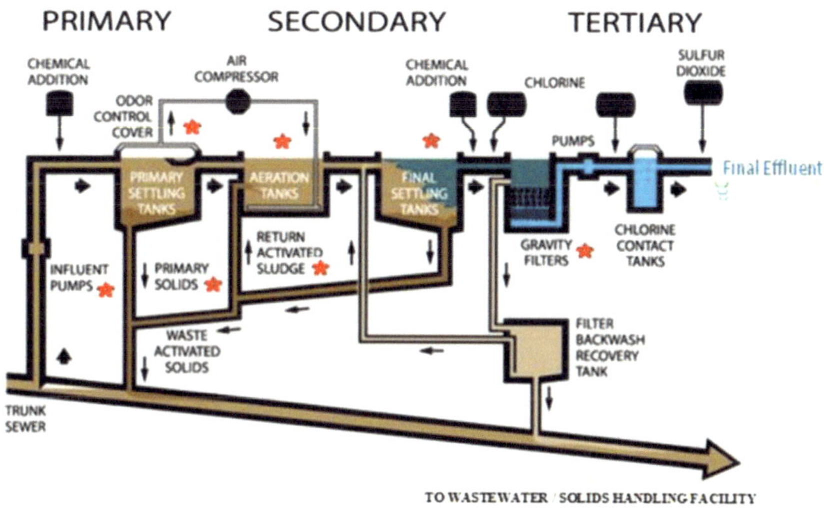

Figure 4. Typical processes of a tertiary WRP. Primary, secondary, and tertiary processes are indicated (modified) [15].

According to Carr et al., most of the microplastics from the influent are removed in the primary treatment zones in which solid skimming and sludge settling processes are applied. In a study published in 2016, effluent discharges from ternary and secondary plants in Southern California were investigated. The importance of understanding the fate and pathways of microplastics in each step of the WWTPs was highlighted [15]. For samples collected in different points along the treatment process, filtration process was applied using an assembled stack of sieves with mesh size between 400 and 45 μm to identify the presence of microplastics in each step. For ternary plants, it was observed that effluent discharge is not a significant source of microplastics. On the other hand, for secondary effluents microplastics were present and can contribute to the contamination of oceans and surface water [15].

Another study conducted in Australia by Ziajahromi investigated the composition of the effluent along primary, secondary, and tertiary treatment. This study detected 0.28, 0.48, and 1.54 microplastics per liter in each respective treatment step. Dayachenko and coworkers [22] observed the presence of 0.09 particles every approximately 4 l of effluent, but this number could reach 0.64 particles in the peak flow of a WWTP in the USA, which uses wet peroxide oxidation. Most of the particles were classified as secondary, and the presence of microbeads accounted for only 10% of the total [17]. Estahbanati and Fahrenfeld investigated the contribution of WWTPs for microplastic pollution in freshwater environment in the USA. In a study of the spatial distribution pattern of MMP in a WWTP, the major presence of secondary microplastics along the treatment process was observed, but the concentration of primary microplastic increased downstream of WWTP was noticed. According to Ziajahromi, although the number of particles does not seem to be quite threatening, when taking into consideration the enormous amount of water that is treated and discharged on daily basis by these facilities, these numbers can represent a very large amount of microplastics entering an aquatic environment. Based on the study of twelve WWTPs in Germany, a study conducted by Mintenig revealed that these facilities can be a sink as well as a source of microplastics which have significant contribution to environment pollution [24].

5. Challenges of preventing microplastic pollution

There are many challenges related to the investigation of microplastic pollution about prevention. Although the interest of the scientific community has been growing toward this subject and the number of published studies has been increasing, the lack of solid definition of what can be considered as microplastic makes it difficult to compare the results obtained in different investigations. Also, there are differences between the methodologies applied on the studies; hence, the results are not always comparable between themselves.

When it comes to the practical application, one of the challenges to prevent microplastic pollution on the water bodies is the lack of technology that effectively retains this kind of material at wastewater facilities. Many researches have been focusing on analyzing the presence of microplastics on the final effluent; the details about the removal of this microplastic in each step of the WWTPs are still unknown. In a previous study, very high removal of microplastics in small capacity wastewater treatment plants was observed [23]. But, even with high removal, small quantities would be discharged into water bodies and would continue causing harmful consequences to the environment.

According to Beljanski et al., there are available technologies that can effectively remove microplastics during wastewater treatment, but they can be expensive, can be difficult to install in existing facilities, and are only used when high-quality standards are required. Membrane bioreactors are an example after primary and secondary treatment, using cross-flow filtration, diffusing only water and small particles. Another drawback of this technology is the high demand for energy and hence higher cost of operation. In a study published in 2016, Beljanski et al. investigated the design of a low-cost, energy-efficient system with easy retrofiltration. Two different filter media were investigated in terms of clogging, retrofiltration capacity, and short-term durability [25].

In Canada, there are many facilities which use membrane technology during the treatment process, but it is not reported that it is a role specifically for microplastic removal. Previous studies in other locations have shown that membrane processes can reduce the amount of microplastic in the final effluent water, but there are many questions about the economic feasibility of its utilization due to the high cost of implementation. In 2016, Michielssen et al. evaluated the efficiency of different unit processes at three WWTPs in removing small anthropogenic little (SAL). The facilities from which the samples were obtained employed either secondary treatment (activated sludge) or tertiary treatment (granular sand filtration) as the final step, as well as a pilot membrane bioreactor system that finishes treatment with microfiltration. It was observed that the membrane bioreactor plant retained a higher percentage of SAL (99.4%) [8].

6. Conclusion

The presence of microplastics is becoming a dangerous environmental concern. Part of the problem stems from the fact that it can be difficult to pinpoint the exact source of the microplastics because of their relatively fragmented nature, small size, and a wide range of potential sources. Microplastics have become a threat to the environment, a concern reflected by sites with unusually high concentrations and a possibility of even greater concentrations in the future. As a consequence, the use and subsequent release of microplastics must be drastically reduced as part of a global initiative even prior to the availability of research studies outlining the long-term risks involved. Monitoring programs can play a key role in the prevention and management of microplastic pollution. The majority of countries, on the other hand, have not developed a strategic approach to researching the primary sources of microplastics accumulating in the water sources or methods of effectually addressing their specific properties. Some nonprofit organizations have committed to collecting and investigating data from various regions in an attempt to track and interrogate these concerns globally, specifically in difficult-to-access or isolated locations. Researchers need to collaborate on an effort that can provide a practical strategy for minimizing applications of microplastics. Although several research projects have examined the effects of microplastics in relation to the final effluent, the particular dynamics involved in the microplastic removal during each separate step of the wastewater treatment plant are still unknown. Thus, it is critical that low-cost and energy-efficient membrane bioreactor systems are designed and implemented, for applications such as primary and secondary treatments at wastewater treatment plants.

Acknowledgements

The authors like to acknowledge the support of the University of Saskatchewan and Ryerson University.

Author details

Heloisa Westphalen[1] and Amira Abdelrasoul[2]*

*Address all correspondence to: amira.abdelrasoul@usask.ca

1 Department of Chemical Engineering, Ryerson University, Toronto, Ontario, Canada

2 Department of Chemical and Biological Engineering, University of Saskatchewan, Saskatoon, Saskatchewan, Canada

References

[1] National Oceanic and Atmosferic Administration. What are microplastics? [Internet] 2016. Available from: http://oceanservice.noaa.gov/facts/microplastics.html [Accessed 2016-12-16]

[2] Ballent A, Corcoran PL, Madden O, Helm PA, Longstaffe FJ. Sources and sinks of microplastics in Canadian Lake Ontario nearshore, tributary and beach sediments. Marine Pollution Bulletin. 2016;**110**(1):383-395. DOI: 10.1016/j.marpolbul.2016.06.037

[3] Wu WM, Yang J, Criddle CS. Microplastics pollution and reduction strategies. Frontiers of Environmental Science & Engineering. 2017;**11**(1):6. DOI: 10.1007/s11783-017-0897-7

[4] Duis K, Coors A. Microplastics in the aquatic and terrestrial environment: Sources (with a specific focus on personal care products), fate and effects. Environmental Sciences Europe. 2016;**28**(1):2. DOI: 10.1186/s12302-015-0069-y

[5] Anderson JC, Park BJ, Palace VP. Microplastics in aquatic environments: Implications for Canadian ecosystems. Environmental Pollution. 2016;**218**:269-280. DOI: 10.1016/j.envpol.2016.06.074

[6] Estahbanati S, Fahrenfeld NL. Influence of wastewater treatment plant discharges on microplastic concentrations in surface water. Chemosphere. 2016;**162**:277-284. DOI: 10.1016/j.chemosphere.2016.07.083

[7] Ivleva NP, Wiesheu AC, Niessner R. Microplastic in aquatic ecosystems. Angewandte Chemie International Edition. 2016;**56**(7):1720-1739. DOI: 10.1002/anie.201606957

[8] Michielssen MR, Michielssen ER, Ni J, Duhaime MB. Fate of microplastics and other small anthropogenic litter (SAL) in wastewater treatment plants depends on unit processes employed. Environmental Science: Water Research & Technology. 2016;**2**(6):1064-1073. DOI: 10.1039/c6ew00207b

[9] Desforges JPW, Galbraith M, Dangerfield N, Ross PS. Widespread distribution of microplastics in subsurface seawater in the NE Pacific Ocean. Marine Pollution Bulletin. 2014;**79**(1):94-99. DOI: 10.1016/j.marpolbul.2013.12.035

[10] Mason SA, Garneau D, Sutton R, Chu Y, Ehmann K, Barnes J, et al. Microplastic pollution is widely detected in US municipal wastewater treatment plant effluent. Environmental Pollution. 2016;**218**:1045-1054. DOI: 10.1016/j.envpol.2016.08.056

[11] Ministry of Food and Drug Safety. MFDS NEWS. Retrieved December 21, 2016, from http://www.mfds.go.kr/eng/eng/index.do;jsessionid=7zAC3OERF5DAcqd6n1OsALGJBEKoLbqPtu9gq4gckArzCD1zgdK5txlFLaPEIlzd?nMenuCode=15&page=1&mode=view&boardSeq=71055

[12] Davidson K, Dudas SE. Microplastic ingestion by wild and cultured Manila clams (*Venerupis philippinarum*). Archives of Environmental Contamination and Toxicology. 2016;**71**(2):147-156. DOI: 10.1007/s00244-016-0286-4

[13] Miranda DD, de Carvalho-Souza GF. Are we eating plastic-ingesting fish?. Marine Pollution Bulletin. 2016;**103**(1):109-14. DOI: 10.1016/j.marpolbul.2015.12.035

[14] Santillo D, Miller K, Johnston P. Microplastics as contaminants in commercially important seafood species. Integrated Environmental Assessment and Management. 2017;**13**(3): 516-521. DOI: 10.1002/ieam.1909

[15] Carr SA, Liu J, Tesoro AG. Transport and fate of microplastic particles in wastewater treatment plants. Water Research. 2016;**91**:174-182. DOI: 10.1016/j.watres.2016.01.002

[16] Magnusson K, Norén F. Screening of microplastic particles in and down-stream a wastewater treatment plant. 2014;**1**(C55):1-22

[17] Ziajahromi S, Neale PA, Rintoul L, Leusch FD. Wastewater treatment plants as a pathway for microplastics: Development of a new approach to sample wastewater-based microplastics. Water Research. 2017;**112**:93-99. DOI: 10.1016/j.watres.2017.01.042

[18] Ontario. (2016, February 23). Microplastics and Microbeads. Retrieved December 20, 2016, from Ontario, https://www.ontario.ca/page/microplastics-and-microbeads

[19] Eriksen M, Mason S, Wilson S, Box C, Zellers A, Edwards W, et al. Microplastic pollution in the surface waters of the Laurentian Great Lakes. Marine Pollution Bulletin. 2013;**77**(1):177-182. DOI: 10.1016/j.marpolbul.2013.10.007

[20] Plastic Microbeads will be Banned in Canada, Effective mid-2018 (2016, November 4) Canada. Retrieved from http://globalnews.ca/news/3047732/plastic-microbeads-will-bebanned-in-canada-effective-mid-2018/?campaign_id=A100

[21] Pescod MB. Wastewater treatment and use in agriculture-FAO irrigation and drainage paper 47. Strategies. 1992;**5**:6

[22] Dyachenko A, Mitchell J, Arsem N. Extraction and identification of microplastic particles from secondary wastewater treatment plant (WWTP) effluent. Analytical Methods. 2017;**9**(9):1412-8. DOI: 10.1039/C6AY02397E

[23] Murphy F, Ewins C, Carbonnier F, Quinn B. Wastewater Treatment Works (WwTW) as a source of microplastics in the aquatic environment. Environmental Science & Technology. 2016;**50**(11):5800-5808. DOI: 10.1021/acs.est.5b05416

[24] Mintenig SM, Int-Veen I, Löder MGJ, Primpke S, Gerdts G. Identification of microplastic in effluents of waste water treatment plants using focal plane array-based micro-Fourier-transform infrared imaging. Water Research. 2017;**108**:365-372. DOI: 10.1016/j.watres.2016.11.015

[25] Beljanski, A. Efficiency and effectiveness of a low-cost, self-cleaning microplastic filtering system for wastewater treatment plants. In: Proceedings of the National Conference On Undergraduate Research (NCUR); 7-9 April 2016; Asheville, North Carolina. 2016. p. 1388-1395

Section 2

Water Management

Chapter 6

Territorial Integration of Water Management in the City

Susana Neto

Additional information is available at the end of the chapter

http://dx.doi.org/10.5772/intechopen.72876

Abstract

This chapter addresses the need for an integrated approach of water management and governance in the urban areas. Water is understood as a natural and social common good, and the relations between different uses and current management practices are analyzed. This approach needs otherwise to influence not only the technical and social spheres but also the policy one. Therefore, a policy effective integration of the comprehensive and interdisciplinary understanding of water problems is advocated and proposed under the concept of integrated urban water policy (IUWP). The final objective is a contribution to an adequate conceptual and operational framework that enables a better and more effective understanding of the multiple dimensions and complexity of integrated management of water uses, in coherent relation with land use planning.

Keywords: integrated water resources management, water governance, territorial integration, integrated urban water policy

1. Introduction

> *"Water connects us. It connects us by flowing across imposed boundaries, by linking diverse terrains and settlements"* (Federico Mayor).[1]

Water is essential to life. The availability of 'good-quality' water is not only the basis of all biological processes but also the maintenance of biodiversity and the functioning of

[1]Keynote address 'Water and Civilization' given by Professor Federico Mayor, Director-General, UNESCO, at the First World Water Forum held in Marrakesh, Morocco, 22 March 1997.

ecosystems. Water has production functions (in terms of biological processes), maintenance functions (in terms of ecosystems) and recovery functions (through the cycles of materials).

Natural ecosystems and farming crops are the great consumers of fresh water. The competition between these consumers has been directly and indirectly intensified by urban land occupation and population growth: directly, as more land for farming means significant change to original ecosystems and, indirectly, because the growing population requires farming to become more productive and to occupy more land.

This occupation not only leads to the change in land uses in question but also to additional pollutants in the soil and associated water masses and the depletion of biological and genetic diversity. In time, the ecosystems become incapable of maintaining their functions of producing water resources with the essential characteristics to sustain human life and activities. This depletion of vital functions performed by the ecosystems endangers the sustainability of human activities like farming, aquaculture and fishing.

The river basin is the natural unit for the occurrence of water resources. River basins constitute large-scale ecosystems that combine land (forest, pasture) and aquatic components (rivers, lakes, wetlands), supplying a great diversity of vegetation and animal habitats. River basins are reference units for the hydrological cycle and the natural resources associated with it, including the filtration and regeneration of elements of the biogeochemical cycle. The main objective of water management is to provide fresh water for human consumption, for food from agriculture and fishing, for hydro-energy, and to all waterside amenities. These territorial units are also the natural link between the land and the sea that can sustain both the movement of fish species and materials between inland and marine waters (or vice versa) and the human need for navigable watercourses. This diversity of functions in the same territory is water's specific virtue as a natural integrator. On the other hand, this virtue also contributes to the difficulties of institutional design for effective water management in several ways [1]. As Giordano and Shah ask, the basin has been always put forward as the key pillar of IWRM implementation and the natural management unit for water resources; there is no evidence that the basin approach is needed [2].

The ecological processes occur and function at various scales—from very small and effecting the proximity to very large with deferred effects in time—which can present inter-scale variations due to the interference of various, often unpredictable, factors. Accordingly, interactions between the different territorial scales condition the end results arising from the referred processes.

When attention is focused exclusively on a specific scale, it can hinder the perception of these interactions, resulting in institutional responses that are not the most suitable because they do not cover the whole process [3]. On the other hand, the issues of water security demand that different scales are considered simultaneously [4]. As referred by Uschi Eid (UN), it is evident that addressing water security problems requires integrating and taking into account different spatial and temporal scales and water governance and management have not been particularly effective in the past in doing so [5].

The basic ideas of integrated water resource management are nearing 100 years of age, being a call to consider water holistically, to manage it across sectors and to ensure wide participation in decision-making [2].

The need for territorial integration is not expressed immediately in current institutional systems and requires various adaptations and articulations between organizations and, above all, between the powers of different levels. On the other hand, the approach of the effects of human occupation on the ecological processes and of the long-term consequences of these alterations on the natural systems, and on the expectations of resource use by human activities, is a complex task for which various fields of knowledge are required. These two factors contribute to a growing need to adapt the operations of the organizations responsible for water management in a way that is not always compatible with the sectoral nature of the existing structures.

2. Water functions

2.1. Water as a multidimensional entity

'In earlier generations, water was seen as a technical issue – a question of organising proper water supply for human societies and agricultural production' [6]. There is now greater recognition that while water has a strong technical component, it is more about managing people, their interaction with the natural environment and the services ecosystems provide.

Due to its cyclic, dynamic and transversal intrinsic nature, water problems have many dimensions and overlap with many scientific disciplines and fields of management, planning and policy.

Modifications of the large-scale ecosystems and natural cycles are having strong impacts on water quality and interfere with the main 'parallel functions' of water. Falkenmark is still very up to date in her definition of the functions for which it is absolutely crucial to create management models that are able to maintain functionality as [6]:

- **Health function**: Safe water is crucial for protecting the survival of a healthy population—this is the perspective of the sanitary engineers.
- **Habitat function**: Aquatic flora and fauna are critically dependent on the characteristics of the water in the water body in which they dwell—this is the perspective of ecologists.
- **Two carrier functions**: Of dissolved material and of eroded material—this is the perspective of hydrochemists and geographers, respectively, to eliminate the pollution source since treatment of waste water is expensive.
- Regional water scarcity in the future will become a limiting factor for **agricultural production**; socio-economic planning will, therefore, have to be adapted to actual water constraints. Inevitably, we will have to develop policy tools capable of managing the shortage of common water resources between competing actors.

- Energy choices in dry climate regions are crucially influenced by water availability because water is required for almost every aspect of **energy production** and use: as a driving medium, as a cooling medium and as an energy transfer medium.

2.2. Water as a human right

'The human right to water is indispensable for leading a life in human dignity. It is a prerequisite for the realization of other rights' [7].

The dimension introduced in the world debate by the United Nations Declaration (*General Commentary no. 15 of the Committee of Economic, Social and Cultural Rights of the United Nations* [7]) has recognized access to drinking water and sanitation as a human right.

Access to water must also be discussed in relation to the scarcity concept, though access to water is a human need and its satisfaction is a political objective.

The European Declaration for Water stated in 2005 that the 'availability of quality water in unlimited quantities 24 hours a day and 365 days a year for multiple uses and at extremely reasonable rates, beyond merely satisfying the human right to a basic share of drinking water, is indeed a conquest of public health, welfare and social cohesion' [8].

In this European level political agreement among scientists, researchers and activists for water rights, it was considered that the 'access to this general interest equity or value must be recognized and guaranteed to all, as citizens' social rights'. Several movements towards this attempt can be referenced in a framework that considers water as an 'essential good' and a 'human right', following the authors Morley [9], Petrella [10] and Sadler [11].

3. Contemporary water challenges

3.1. Water management: global changes and increasing complexity

Until the 1970s, there were essentially isolated incidences of environmental problems (accumulation of pollutant residue, infected waters, polluted atmosphere, etc.) that have had more or less limited effects and generally restricted to the territory of origin. Since the 1980s, and in more particularly the late 1980s and early 1990s, environmental problems have become increasingly complex. The effects started to be cumulative and could not be explained by a single phenomenon or cause, about which there were varying degrees of knowledge, and problems started to acquire greater spatial scope. It therefore became indispensable to adopt a comprehensive perspective for analysis and intervention in order to understand and address phenomena like the greenhouse gas effect, the rarefaction of the ozone layer or climate change.

The nature of environmental problems evolved, forming combinations of symptoms and factors that required ever-increasing scrutiny. In the 1990s, there was believed to be three broad groups of problems [12]:

- Generalized degradation of soil resources (loss of roughly 1.2 billion hectares of land in the previous 50 years).
- Growing accumulation of gases in the atmosphere, responsible for the worsening of the greenhouse gas effect and climate change (in particular CO_2, the emissions of which would have to decline about 60% at that time).
- Biodiversity increasingly at risk (clearing forests, drainage of humid zones, intensification of the harvesting of living resources, degradation of ecosystems that resulted in the extinction of thousands of species).

In this context, water quality constitutes a central theme for policy action. The concept of water quality has evolved insofar as recognition is given to the growing interactions and levels of complexity in the degradation processes of water resources. These include not only the cumulative factors and effects in space and time but also the persistence of those effects caused and the difficulty in taking isolated localized actions.

In the first decade of the twenty-first century, the globalization of the problems has become more evident. Global changes like climate changes, demographic changes and economic crisis are affecting directly and indirectly the resources in water and their natural ecosystems, with growing effects at transnational and world scale, demanding international concerted action.

The last 5 years of the twentieth century were characterized by an overall tendency of continuous glacier melting. This decline will have impacts on both the sustainability of the water resources in basins, which depend on glaciers and on their ecosystems [13]. However, demographics and the increasing consumption that comes with rising per capita incomes are the most important drivers or pressure on water [14].

The world's population is growing by about 80 million people a year, implying an increased freshwater demand of about 64 billion cubic meters a year. Most population growth will occur in developing countries, mainly in regions that are already experiencing water stress and in areas with limited access to safe drinking water and adequate sanitation facilities [14].

Human population growth and the expansion of economic activities are collectively placing huge demands on coastal and freshwater ecosystems. The increase in the number of people without access to water and sanitation in urban areas is directly related to the rapid growth of slum populations in the developing world and the inability (or unwillingness) of local and national governments to provide adequate water and sanitation facilities in these communities. The world's slum population, which is expected to reach nearly 900 million by 2020, is also more vulnerable to the impacts of extreme weather events. It is however possible to improve performance of urban water supply systems while continuing to expand the system and addressing the needs of the poor [5].

The management of scarce resources places new challenges to the institutional and organizational systems and schemes that are operating in a traditional way. This usually relates to continuing a 'water supply focus' of increasing the levels of services and investments in infrastructures in response to the increasing needs imposed by population growth, urbanization and intensification of irrigated areas. The need to manage the 'demand', increasing awareness of water scarcity, equity and sustainable uses of the available resources is still finding institutional and operational answers, though the principles are generally accepted by the agents of management and planning. This will mean a shift towards a new paradigm in water management, search for innovative and sustainable solutions, alternative sources of water (like more recycling and rainwater collection) and a new institutional framework, where the sectors may cooperate effectively and the public administration can address complex problems through new and flexible learning organization and structures.

It is still a fact that, as stated by Falkenmark almost 20 years ago, societal rules manifested in legislation and administration are *quite inconsistent with natural laws* [6]. Even though the same water is used for a whole set of different uses while running down the river basin, these uses are administered by different authorities as if they were not connected.

The complexity involved in water resource management and its linkages to land use adds even more difficulties to the management systems and to the decision-makers. Water management is increasingly complex and dynamic, requiring more flexible and adaptive responses.

3.2. Access to water, transparency and allocation of resources

The effects of water-depleting and water-polluting activities on human and ecosystem health remain largely unreported or difficult to measure [14], with growing need for a stronger and effective protection of ecosystems and the goods and services they produce.

Figure 1 shows project global water scarcity in 2025. The scarcity of water does not only refer to human needs. Over-use and over-allocation of water resources for human activities can lead to substantial reductions in the flow of watercourses necessary to maintain associated water-dependent ecosystems. This scarcity causes a reduction in the capacity of fresh water ecosystems to provide all the environmental services and can result in their irreversible degradation and that of the species depending on them. On the other hand, rising pressure from increasing population will endanger the maintenance of minimum environmental requirements. This situation is aggravated when growth is combined with the natural less favorable conditions and variability of resources.

Therefore, environment (including aquatic environment) should not be seen as complements to economic development but the foundation (and limits) on which development is built. Reflection should focus on the interdependence between the natural means and social systems. Water should be considered not as a resource which is available in varying degrees, but as a complex natural and integrating means, and our survival in this planet depends on its interdependence with other systems.

Ten years ago (2006) 13% of the world's population did not have access to enough food to live a healthy and productive life, yet the ability, technology and resources needed to produce enough food for every man, woman and child in the world do currently exist. Lack of health, financial or natural resources such as land and water and lack of skills to link productive activities with remote markets and ensure employment are all intimately related to poverty [13].

http://dx.doi.org/10.5772/intechopen.72876

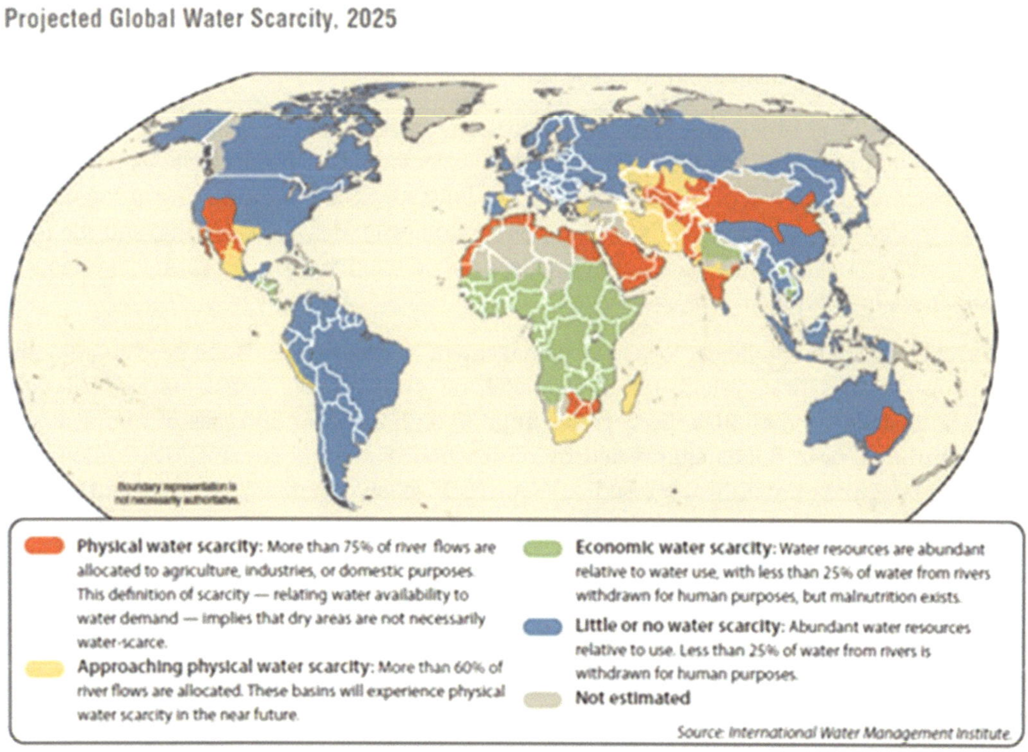

Figure 1. Projected global water scarcity in 2025 (source: IWMI 2000).

By 2050, agriculture will need to produce 60% more food globally and 100% more in developing countries. As the current growth rates of global agricultural water demand are unsustainable, the sector will need to increase its water use efficiency by reducing water losses and increase crop productivity with respect to water [5].

3.3. Water for urban areas

By 2030, the number of urban dwellers is expected to be about 1.8 billion more than in 2005 and to constitute about 60% of the world's population. Ninety percent of the increase in urban population is expected in developing countries, especially Africa and Asia, where the urban population is projected to double between 2000 and 2030. Coastal areas, with 18 of the world's 27 megacities (greater than 10 million), are thought to face the greatest migration pressure. *The net implication is that the world will have substantially more people in vulnerable urban and coastal areas in the next 20 years* [14].

In 2000, more than 900 million urban dwellers (nearly a third of all urban dwellers worldwide) lived in slums. A slum dweller may only have 5 to 10 liters per day at his or her disposal compared to the UN suggested minimum requirement of 20–50 liters. A middle- or high-income household in the same city, however, may use some 50–150 liters per day, if not more [13].

As urban populations increase and local surface and groundwater sources are depleted or polluted, many major cities have had to draw freshwater from increasingly distant watersheds as to meet their rising demand for water [13].

Water occurrence in urban areas has the particularity of being either 'natural' or 'artificially' produced. Its presence takes the form of natural water bodies that constitute esthetical and leisure resources, the form of rainfall, flowing over built surfaces and infiltrating in subsoil; the form of treated water conducted through pipes for domestic uses and drinking; and the form of residual water in various degrees of degradation, conducted to sewage treatment systems or directly back to natural streams or the sea.

The specificity of the challenge for water management in urban areas is the need to integrate the natural hydrological systems functioning and the artificially build systems, in a balanced way and in order to minimize the negative impacts on both sides (increase of runoff due to paved surfaces, flash floods aggravated by obstruction of natural streams, overexploitation and increasing pressure on water resources sources to satisfy increasing demand and urbanization growth, contamination of groundwater's, surface and coastal waters by the waste waters and the urban solid wastes and all kinds of particulates generated in urban areas, etc.).

The use of unexploited resources like storage of rainwater in buildings and open places (gardens, parking areas) is one very relevant action to take seriously inside urban areas and must be considered as a top priority in any management plan, not only in dry or semidry regions.

4. Integrated water management, water governance and policy change

4.1. Achieving synergies between water and other policies

Emerging first at Mar del Plata and evolving through subsequent water fora was recognition of the need to integrate all aspects and dimensions of the water cycle, to achieve integrated water resource management (IWRM). More recently, there is increasing recognition that operating only within the water domain is not sufficient as the greatest factors influencing water management are beyond water policy in *stricto sensus*, in areas such as national economic policy.

Following the call for a more holistic approach to water resource management, water utilities and land-use planners need also to coordinate to overcome these water resource challenges in the urban areas [15]. Planners and utilities need to work together to implement more green infrastructure to better manage storm water, more onsite reuse to reduce potable demand and more green buildings to reduce potable demand and better manage water within buildings [15].

Agreeing with Engle that IWRM is fundamentally about governance arrangements, we also note that IWRM stresses the interconnectedness of catchments and users; there is no universal model for the way in which management institutions are structured and linked [16].

The need to improve assessment of the effectiveness of water governance and its impact on the implementation of IWM in an integrated way with territorial planning in urban areas relates to

http://dx.doi.org/10.5772/intechopen.72876

the responsibility of people and institutions involved in water management and urban planning, in any country, to determine and prioritize actions necessary to improve water and territorial governance. The central question therefore revolves around the premise that water should not be managed as a mere resource but as a complex natural entity, whose frontiers extend to environmental, institutional and social (including the economic aspects) spheres and whose reference is always territorial. This premise results not only from the specific characteristics of the water cycle, closely associated with the physical territory (including the soil and subsoil, vegetation cover, ecosystems, atmosphere and climatic factors). It also results from the fact that the human communities have occupied and transformed this territory, from the beginnings of sedentary societies and the time when man learnt to store water, interfering with this cycle and being also conditioned by it.

Consequently, a 'territorial' approach shall consider all the means in which water is manifested and in which it remains or contributes to the various biological, geological and chemical cycles, as well as all the occurrence, draining and natural infiltration processes on one hand and all the reserve, abstraction, consumption, change of water courses and/or physical-chemical and biological conditions on the other. This approach does not exclude social relations of cooperation or conflict towards access to water, in particular regarding the occupation and changing processes of the use of land and territory.

4.2. Water functions and territorial scales

Falkenmark has deeply developed the analysis of the water and land use interaction. Her studies on the articulation of the cyclical support functions of the materials by the hydrological cycle, through the physical territories of the river basins, provide greater understanding of the interactions between the water functions and the physical territory, as illustrated in **Figure 2**.

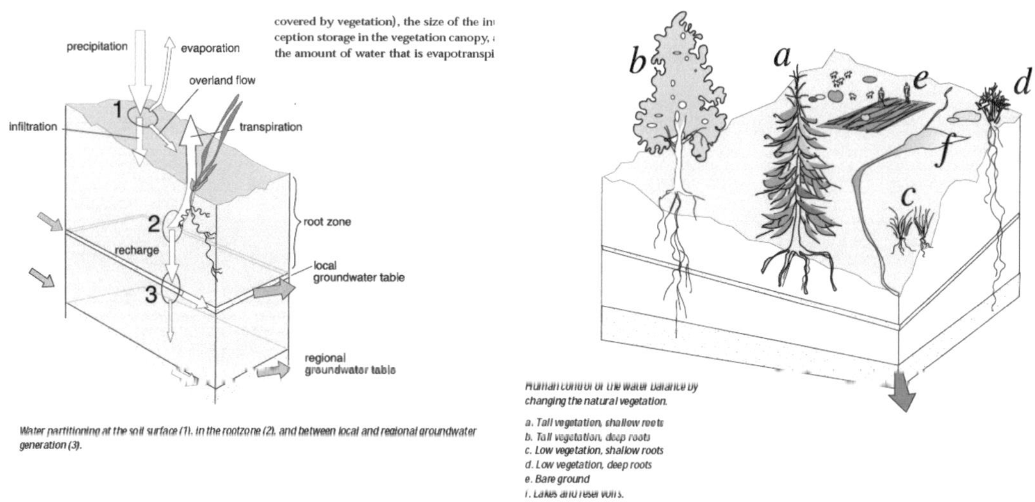

Figure 2. Water-soil interactions (source: Falkenmark [6]).

Falkenmark discusses the different administrative divisions and the need to consider flows of water imports from upstream to downstream through these jurisdictions. The buffering capacity of ecosystem services (flood regulation, drought mitigation, wetland water storage, preparing room for the river, clearing invasive trees, etc.) is essential, and these dimensions need to be considered in a holistic management approach.

Figure 3 presents the division of a basin in three regions and the resulting precipitation—evapotranspiration and its circulation to the surface or through the subsoil of regions from upstream to downstream.

4.3. Territorial integration of water management

The conclusions of the different *fora* and international committees on worldwide water management give ever-greater priority to the recognition that water is not a 'mere resource' but a complex and multidimensional entity. They highlight the need for concerted action between the different actors (government and society, stakeholders, etc.), in order to respond to the need for more 'effective' water governance [1].

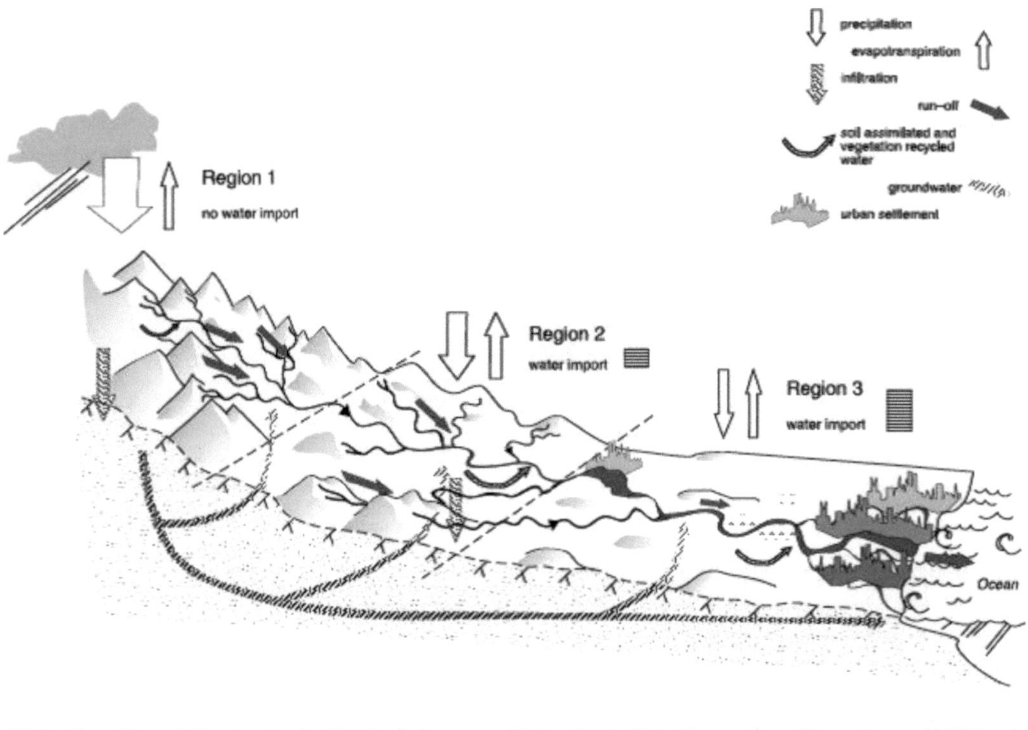

Water flows through the river basin. This basin is assumed to be divided into three regions. The water availability of each region is determined by the local precipitation and evapotranspiration as well as the import of water from upstream regions. The rain water that falls on a region and which does not disappear as evapotranspiration generates runoff.

Figure 3. Water flow through the river basin territory (source: Falkenmark [6]).

Areas of further research for better operational results from integrated water management practices and for improving water governance can be identified [1].

- Coherent articulation of spatial plans aiming at the integration of land uses, water and sustainability.
- The role of state agencies aiming at integration and participation.
- Enhancing opportunities for local solutions
- Capacity building among planners aiming at adequate response to new challenges, adequate knowledge concerning contemporary planning problems (participation, governance, climatic change, etc.).

A river basin or drainage basin is a territory that contains a watercourse and its tributaries. The river basin should be the starting point for the analysis of the water resources, associated ecosystems and habitats that depend on them, as well as the analysis of all the interdependencies resulting from the scale(s) of occurrence from natural processes or the dysfunction's triggered by human occupation and activities.

However, to effectively integrate water resource management with other natural resources, we need to think beyond the basin, and planning at any level always involves the public administration and the elected representatives of the state. Good government requires adequate networking and the integration of all sectors in planning activities. As quoted before, *the UN recognizes that drivers and policies outside the water sector often have more impact on water management than those within* [17].

Territorial integration of different sectoral strategies is the only way to match objectives at each scale of analysis. This integration demands an approach to water problems relating to all sectors, taking into account the other scales of analysis, beyond the river basin.

4.4. Water policy and the institutional systems

Decisions on the appropriate jurisdictional level for water resource allocation and management need to take the impact of all sectors at all scales into account. Furthermore, the level for jurisdiction must also recognize the specific role of the state in defining the priorities for the use of the available water. The promotion of good 'governance' in addition to good government depends primarily on the political will of the government and public administration bodies. Participation of stakeholders and sectors, as well as end users of water infrastructures and water amenities, is a central issue in governance. The need for integrated representation of all relevant human settlements, which are not always wholly within one basin, is also a key issue for good water governance [6, 18, 19, 1].

Water policy is a broad field where new perspectives for water management and planning can be framed. Le Meur defines this distinction between water management and 'water politics', based on the more comprehensive perspective and the inclusion of the social dimension, as the most relevant difference. Water politics should replace water management in order to involve diverse social actors and groups and follow an approach that goes apart from the managerial perspective that the 'right solution' (technical and institutional) exists and can be worked out with expertise and participation [20].

Allocation of available resources, according to priorities defined by the state and the control of these resources, is a prerogative of the state action, through its policies. This responsibility of how to deploy a resource to the national advantage is the key to water governance at the beginning of the twenty-first century, accordingly, and it is 'how, through politics, the State can achieve this fairly and equitably, without reducing incentives for efficient use of the resource' [20].

Fragmentation at administration levels is a traditional problem in environment and water institutions, with the different authorities handling different issues so that it becomes more difficult to allow interdependencies.

Even though reality is multidimensional and complex, all our institutional organizations are oriented to unidirectional and unidimensional interventions, leading to the dilemma stressed by Falkenmark, and discussed previously. It also happens that individual institutions have specific and often opposing objectives [6, 18]. The legislative framework is built to frame action in compartmentalized 'worlds', often hindering any interaction and cooperation that would be crucial to address these complex nature problems.

The 'land/water dichotomy characterized by a mental image of land as opposed to water' that was present at the time (1999) is still a reality, with policy structure based on the separation between land and water as entities to be managed in different ways and by different management systems. What usually happens in Europe, and many other regions in the world, is that land and water are administratively linked mainly through the use of environmental impact assessments in the case where land use changes produce side effects on water, not going beyond that legal guiding framework.

Consideration of the **territorial character of water** represents a complex challenge for water management but unavoidable in the urban areas. While cities are impacted by global changes, some urban structures become more and more static in comparison to the unexpected and faster changing contexts [21]. Accepting also that urban settlements incorporate an increasing complexity of interlinks and interdependencies within their economic and social tissue, effective solutions will not arise exclusively through 'traditional, institutional or formal' approaches. Therefore, *out-of-the-box*, experimental and cross scale responses may be necessary [21].

5. Conclusions

There is a crucial need for better understanding of the extended urban water cycle within the river basin processes at regional level. River basins are the territorial units for hydrological reference and for understanding the natural cycle in all its dimensions, including the links with other natural resources. Integrated water management calls for a territorial integration of these dimensions and the processes.

The diversity of functions in the same territory is one of water's specific virtues as a natural integrator. On the other hand, this virtue also contributes to the difficulties of institutional design for effective water management in several ways. The need for territorial integration is not automatically reflected in the current institutional frameworks, and existing processes require several levels of adaptation, as well as articulation between organizations. On the other hand, the approach of the effects of human occupation on the ecological processes and of the long-term consequences of these alterations on the natural systems, and on the expectations of resource use by human activities, is a complex task for which various fields of knowledge are required. These two factors contribute to a growing need to adapt the operations of the organizations responsible for water management in a way that is not always compatible with the sectoral nature of the existing structures [1, 22].

There is a clear call for better integrated land and water management in the urban sphere by developing analytical frameworks and operational tools to assess the climate change impacts and the role of territorial planning in the urban water cycle. A new conceptual framework based on 'territorial integration', 'sustainable urban development' and 'integrated water management' was proposed to address the new and more complex challenges: an 'integrated urban water management policy' [22]. This approach is based on previous conceptual discussion [6, 19, 1, 23–25].

This contemporary approach aims to provide the policy making sphere with more operational and comprehensive tools to address the multidimensional challenges imposed by urban water management in the context of the current global changes and challenges.

Acknowledgements

This chapter is based on my PhD thesis presented at the University of Lisbon in 2010 [1] and also reflects more recent work [22].

Author details

Susana Neto[1,2]*

*Address all correspondence to: susana.neto@netcabo.pt

1 University of Lisbon, Lisbon, Portugal

2 University of Western Australia, Perth, WA, Australia

References

[1] Neto S. Water, Territory and Planning—Contemporary Challenges: Towards a Territorial Integration of Water Management. Lisbon: Technical University of Lisbon; 2010

[2] Giordano M, Shah T. From IWRM back to integrated water resources management. International Journal of Water Resources Development. 2014;**30**(3):364-376. DOI: 10.1080/07900627.2013.851521

[3] Alcamo et al. Ecosystems and Human Well-Being: A Framework for Assessment–Millennium Ecosystem Assessment (MA). Paris: UNESCO; 2003

[4] Wostl C, Bhaduri A, Gupta J. Water security: A popular but contested concept. In: Handbook of Water Security. UK: Edward Elgar Publishing Limited and USA: Edward Publishing, Inc.; 2015 (this book is available electronically in the social and political science subject collection). DOI: 10.4337/9781782548010

[5] UN/WWAP (United Nations World Water Assessment Programme). Water for a Sustainable World. Report 5. Paris: UNESCO and London: Earthscan; 2015

[6] Falkenmark M, Andersson L, Castensson R, Sundblad K. Water, a Reflection of Land Use: Options for Counteracting Land and Water Mismanagement. In: Falkenmark M, Andersson L, Castensson R, Sundblad K, editors. Stockholm: Swedish Natural Science Research Council; 1999

[7] United Nations/ECE. General Comment on the Right to Water. United Nations Committee on Economic, Social and Cultural Rights (part of the UN Commission on Human Rights); November 2002; published 2003

[8] Arrojo P, Dionysis A, Barraque B, Bressers H, Estaban C. European Declaration for a New Water Culture. Madrid: Fundación Nueva Cultura del Aqua; 2005

[9] Morley D, editor. Perspectives on Freshwater: Issues and Recommendations of NGO's. Report NGO's and Water. London: UNED Forum; 2000

[10] Petrella R. Le Manifeste de l'Eau – pour un Contrat Mondial. Bruxelles: LABOR; 1998

[11] Sadler B, editor. Communication strategies for heightening awareness of water. Report 2, International Hydrological Programme (IHP II project C1). In Studies and Reports in Hydrology. Paris: UNESCO; 1987

[12] World Resources Institute. Global Biodiversity Strategy: Policy Makers Guide. World Resources Institute, The World Conservation Union, United Nations Environment Program in Consultation with Food and Agriculture Organisation (FAO) and UNESCO. Washington, DC: World Resources Institute; 2002, 1992

[13] UN/WWAP (United Nations/World Water Assessment Programme). 2nd UN World Water Development Report: Water, a Shared Responsibility. Paris and New York: UNESCO (United Nations Educational, Scientific and Cultural Organisation) and Berghahn Books; 2006

[14] UN/WWAP (United Nations World Water Assessment Programme). The United Nations World Water Development. Report 3: Water in a Changing World. Paris: UNESCO and London: Earthscan; 2009

[15] Whitler JY, Warner J. Integrated Urban Water Management for Planners–PAS Memo. Water Research Foundation. Chicago: American Planning Association; Published online September/October 2014

[16] Engle NL, Johns OR, Lemos M, Nelson DR. Integrated and adaptive management of water resources: Tensions, legacies, and the next best thing. Ecology and Society. 2011; **16**(1):19. [online] URL: http://www.ecologyandsociety.org/vol16/iss1/art19

[17] UNWater. Climate Change Adaptation is Mainly about Water. UNWater; 2010. http://www.unwater.org

[18] Falkenmark M, Jägerskog A, Schneider K. Overcoming the land–water disconnect in water-scarce regions: Time for IWRM to go contemporary. International Journal of Water Resources Development. 2014;**30**(3):391-408. DOI: 10.1080/07900627.2014.897157

[19] Neto S. Territorial approach in water uses planning: A strategic view. Presented at the expert meeting on Water, Land, Agriculture and Policies in the Mediterranean. Brussels: IPTS/CE, s.n..; 2001

[20] Le Meur PY, Hauswirth D, Leurent T, Lienhard T. The Local Politics of Land and Water: Case Studies from the Mekong Delta. Coll. Études et Travaux, série en ligne n°4, Éditions du Gret, 2005;**4**. www.gret.org

[21] Henriques C, Neto S. Hybrid solutions to address complexity—water governance facing global changes impacts in urban areas. Paper presented to the Conference Adapting to Climate Change Water, Waste and Other Local Infrastructure; 21-22 June; Lisbon, Portugal

[22] Neto S. Water governance in an urban age. Utilities Policy. December 2016;**43**(Part A): 32-41. DOI: 10.1016/j.jup.2016.05004

[23] Brown RR, Farrelly MA. Delivering sustainable urban water management. Water Science & Technology—WST. 5 (SUWM). 2009. p. 8

[24] Mitchell VG. Applying integrated urban water management concepts: A review of Australian experience. Environmental Management. 2006;**37**(5):589-605

[25] LeRoy Poff N, Brown CM, Grantham TE, Matthews JH, Palmer MA, Spence CM, Wilby RL, Haasnoot M, Mendoza GF, Dominique KC, Baeza A. Sustainable water management under future uncertainty with eco-engineering decision scaling. Nature Climate Change. 2015;**14**:09

Urban Development with the Constraint of Water Resources: A Case Study of Gansu Section of Western Longhai-Lanxin Economic Zone

Gao Xiang

Additional information is available at the end of the chapter

http://dx.doi.org/10.5772/intechopen.70850

Abstract

Water is the essential resource for urban development of Gansu section of West Longhai-Lanxin economic zone, which is not only the "Golden Development Line" of Gansu province but also the significant component of the new Silk Road within China. Based on more than 2000 data points reflecting various meaningful aspects of urban systems and water resource systems, and using a quantitative measurement model and ArcGIS, this study aimed at discussing the temporal-spatial variations of water resources constraint on urbanization in the Gansu section. From 1989 to 2007, the water resources constraint intensities (WRCIs) of Gansu section and its nine cities have been generally decreasing, albeit with much fluctuation, and the decrease has been more rapid since 2000, with Lanzhou and Jinchang as the most representative cities. From the perspective of water resource constraint on urbanization, the research on urbanization process of Gansu section is not only necessary for the shaping of an independent theoretical system on relationship between water resource and urbanization, but also has very crucial practical significance for promoting construction of resource-saving and environment-friendly cities in Gansu section, arid and semi-arid areas as well as for promoting harmonious regional development.

Keywords: Gansu section of West Longhai-Lanxin economic zone, silk road, water resource constraint, urbanization, temporal–spatial variation

1. Introduction

Water is the strategic resource and basic material for human being's survival and for the economic and social development. Biswas has made special researches on the world development model, including the city, and holds opinions that overall developing trend of urban economic activities around the world will face the rigid constraints of increasing resource shortage in the twenty-first century, including the constraint of water resource [1]. Mcmichael also points out that urbanization will threaten human being's living environment and health in an important way. The urban expansion, industrial growth and population growth will bring large pressures on local water resources and ecological environment [2]. On World Water Day in 2002, Koichiro Matsuura, the Director-General of UNESCO, remarked that "the impending water resource crisis poses one of the most serious challenge to the current world". Changming investigated and believed that water consumption over the whole world in the nineteenth century increased by 8 times, among which agricultural water increased by 7 times, industrial water by 20 times, urban domestic water by 12 times, and the water consumption quantity doubled every 15 years [3]. And in the twentieth century, world population grew by two times, while the human water consumption still increased by five times on condition that some developed countries had already realized zero growth of water demand. It is predicted that in 2050, urban water use will be equal to the current global water consumption and over 55% population around the world will face the problem of water crisis [4]. Water security problem in the twenty-first century not only directly threatens human being's food security, economic security, and ecological security, but also endangers social security and national security, and even people's living security, especially in arid and semi-arid areas.

This situation brings increasingly difficult choices to China, which is facing water shortage, overall eco-environment deterioration, and the significantly accelerating urbanization. Nobel economics laureate Stiglitz (2000) pointed out that "China's urbanization is a global event with the most profound influence in the twenty-first century other than high-technology". Since 1998, the urbanization level of China has been growing with a rate of 1.5–2.2% points each year. Until the end of 2016, the urbanization rate has reached 57.35%, being the fast urbanization period. The large-scale and fast urbanization in China will highlight the imbalance of water supply and demand, which has become a bottleneck restraining the urban development of our country [5]. Especially in arid and semi-arid areas in northwestern parts such as the Gansu section of western Longhai-Lanxin economic zone, water resource is not only an endogenous variable for urbanization and the social and economic development, but also has an important exogenetic force. In the process of industrialization and urbanization, cities with a fragile eco-environment as the developing background have suffered multi-layers of threats posed by water resources and surrounding eco-environment. However, urbanization is the main approach for these regions to realize modernization, and their urbanization level still has large climbing space compared with that of eastern coastal areas. Therefore, it is especially important and extremely urgent to well coordinate the relationship between urbanization and water resource in these regions. Research on urbanization process of Gansu section from the perspective of water system constraint is not only necessary for the shaping of an independent theoretical system on relationship between water resource and urbanization, but also has very crucial practical significance for promoting construction of resource-saving and environment-friendly cities in Gansu section, arid and semi-arid areas as well as for promoting harmonious regional development.

2. Study area

Western Longhai-Lanxin economic zone starts from Tongguan, Shaanxi province in the east to Urumqi, Xinjiang Province in the west, with the western part of Longhai Railway, the Lanxin Railway, and the high-level roads and communication lines of the same direction as its axis, the provincial capitals and prefecture cities expanding like beads as its nodes, and the vast surrounding rural areas as hinterlands. As a result, an economic region within the scope of 100–150 km in width and over 2700 km in span is formed [6]. On principle of "line links points, points link areas", relying on this trunk transportation line to develop node cities is one of the keys for develop-the-west strategy and also a significant component of the new Silk Road within China.

The Gansu section of western Longhai-Lanxin economic zone (hereinafter referred to as "Gansu section") is an important component of this economic zone, which spans 1542 km and accounts for 57.1% of the zone. *Development Plan on Gansu Section of Western Longhai-Lanxin Economic Zone (2003)* issued by Gansu provincial government shows that Gansu section is composed of nine cities (Tianshui, Lanzhou, Dingxi, Baiyin, Wuwei, Jinchang, Zhangye, Jiuquan and Jiayuguan) as core districts and the non-city influenced districts which are featured with rapid urbanization, high-level openness, strong economy and great developing potential, that are, 42 districts. Directly influenced districts are Pingliang city, Linxia prefecture, Li county, Xihe, Cheng county, Hui county and Liangdang county; indirectly influenced districts are Qingyang city, Gannan prefecture and Dangchang, Wudu, Kang county and Wen county (**Figure 1**).

Gansu section is located at convergence zone of Loess Plateau, Qinghai-Tibet Plateau, and Mongolian Plateau. It is a mountainous plateau landform with complex terrains on which mountain ranges crisscross and various landforms such as high mountain, basin, plain, desert and Gobi deserts with greatly differing altitudes, mainly composed of Hexi Corridor, Qilian Mountains, regions north of Hexi Corridor, Longzhong loess plateau and Longdong loess plateau. Gansu section contains six climatic regions featuring semi-arid climate while the continental climate and drought degree increase from both sides to center areas, making Hexi Corridor become a section with the least precipitation and the most obvious continental climate within this district. Corresponding to drought climate, most districts of Gansu section are with sparse river-network, small runoff, and water resources shortage.

Gansu section is the axial region for economic development of Gansu province. Up to the end of 2015, population in Gansu section reaches 16.359 million, covering 62.93% of total provincial population and concentrating 74.27% of non-agriculture population of the whole province; 2015-year-end agricultural area is 2166,484 ha, covering 66.0% of the province; regional GDP is RMB 514.004 billion, covering 75.70% of the province and concentrating 80.3% of total provincial industrial output value, 84.1% of total construction output value and 83.07% of total wholesale and retail output value. Gansu section is the "Golden Line" in development of Gansu province.

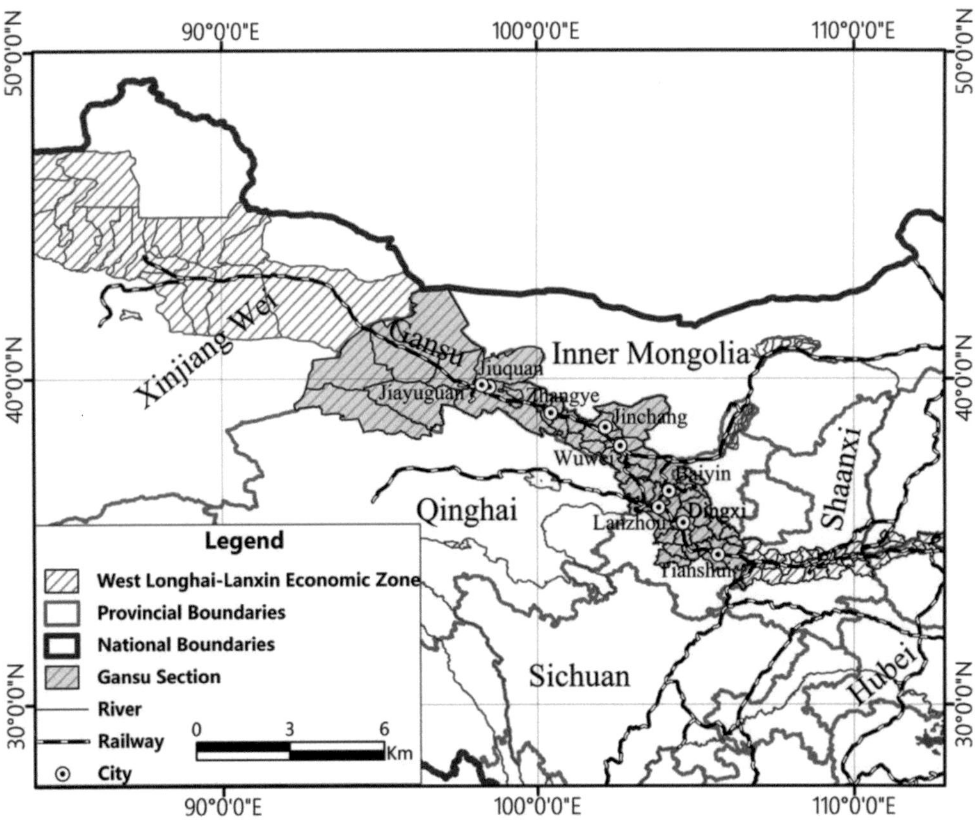

Figure 1. Geographic location map of Gansu section.

3. Materials and methods

3.1. Indicators

The relationship between water resources and urbanization is extremely complex, and no singular indicator can truly reflect the interaction mechanism between the two. Therefore, following available scientific principles and taking into consideration regional dynamics, we first obtained a general set of indicators based on frequency analysis, theoretical analysis and specialist consultation; then, taking into account information overlap, we passed these indicators through correlation coefficient and principal component analysis screening, thereby obtaining final indicators (**Table 1**).

3.2. Data sources and preprocessing

Considering data authoritativeness and availability, the sources of initial data mainly include: precipitation monitoring data from main meteorological stations of Gansu province (1956–2001 Gansu Meteorological Bureau), *Gazette of Water Resources of Gansu Province (1997–2008)*,

	Functional groups	Index
Urban system	Population urbanization	Urban population growth rate, proportion of non-agricultural population, population density
	Demographic urbanization	Proportion of non-agricultural industries, per capita GDP, per capita GDP growth rate, economic density of built-up areas, non-agricultural industries density, proportion of industry added value, industry added value per capita
	Social urbanization	Disposable income, per capita spending, per capita road area, Engel coefficient, no. of graduates (m), no. of doctors (m), no. of telephones (m), proportion of science and education funding, per capita living space
	Spatial urbanization	Per capita built-up area, built-up area density, urban density, urban population density
Water resources system	Water resources background conditions	Per capita water resources, resources equivalent runoff, water production module, drinking water quality compliance rate
	Development level of water resources	Per capita water consumption, resource utilization rate, groundwater development and utilization rate
	Efficiency of water resources development	water consumption (10T CNY GDP), gross irrigation water quota, unit food production water consumption, water consumption of industry added value (10T CNY), rate of industrial water usage repetition, citizen water quota, rural residents water quota, proportion of water saving irrigation area
	Water resources management ability	Hydro-science personal (10T), ecosystem water consumption rate (10T CNY GDP), waste water emission amount, urban sewage treatment rate

Table 1. Indicators of water resource and urban systems.

"Statistical Information on Water Resources of Gansu Province" (1956–2001 Gansu Province Hydrology and Water Resources Investigation Bureau), Gazette of Environmental Quality of Gansu Province (2001–2008), Statistical Yearbook of Gansu Province (1990–2008), City Yearbook of Gansu Province (2001–2008), China City Yearbook (1989–2007), statistical yearbooks of each city for many years, plan for national economy and social development of each city, *Study on Water Availability of Main Functional Districts in Gansu Province, Industry Water-use Quota in Gansu Province.* Some additional data belongs to the data accumulated from project undertaken in the past, like the key project of National Natural Science Foundation—The Contemporary Living Environment and Social Foundation Sustainable Development in Hexi Area.

Based on the initial data we selected more than 2000 valued data points between 1989 and 2007, reflecting various aspects of water resource systems and urban systems. Because of the variations in the dimensions of the initial data for the indicators to facilitate horizontal comparison among the nine cities, we used poor standardization methods to treat the data and make it dimensionless:

$$Z_{ij} = \left(X_{ij} - \min X_{ij}\right) / \left(\max_i X_{ij} - \min_i X_{ij}\right) \tag{1}$$

where Z_{ij} is the attribute value of a target and $\max_i X_{ij}$, $\min_i X_{ij}$ are the maximum and minimum values of a target.

3.3. Methods

The model for the comprehensive measurement of water resources constraint on urbanization is built upon the integrated indicator system of water resources and urbanization, using the entropy technique of supported analytic hierarchy process (AHP) to calculate the weight of indicators at all levels, using a fuzzy membership function model to calculate the specific membership values of each indicator, the weighted average method to individually calculate the integrated indexes of water resource development potential and the indexes of urban synthesized development, and then using these to calculate the water resources constraints on urbanization, and grade their intensity accordingly [7–9].

(1) Using entropy techniques to correct the weight coefficient determined by the AHP method.

Because there is a complex nonlinear relationship between each factor within the water resources and urban systems, we used the entropy technique of supported AHP to determine the weight of each eligible index.

$$\alpha_j = v_j p_j / \left(\sum_{j=1}^{n} v_j p_j \right), v_j = d_j / \sum_{j=1}^{n} d_j, d_j = 1 - \lambda_j, \lambda_j = -\sum_{j=1}^{n} r_{ij} \ln r_{ij} \quad (2)$$

where a_j is the index weight determined by the entropy technique of supported AHP; p_j is the index weight determined by the AHP; v_j is the index information weight; λ_j is the indicator output entropy power and r_{ij} is the standard matrix after normalization using the AHP determination matrix. Using these equations increased the reliability of the results for each factor.

(2) Using a fuzzy membership model to calculate the membership values of each specific indicator.

In order to solve the problem of differing dimensions for different indicators, with consequent summarization difficulties, it is necessary to eliminate the dimension of each indicator. Considering that there are both positive and negative indicators in the system, the 'good' and 'bad' between indicators are to a large extent fuzzy; hence a fuzzy membership model function method was used to quantify the 'value' of each indicator, treating positive indicators with a higher semi-trapezoid fuzzy membership function, as follows:

$$\Phi_{(e_{ij})} = \frac{e_{ij} - m_{ij}}{M_{ij} - m_{ij}} = \begin{cases} 1 & e_{ij} \geq M_{ij} \\ \dfrac{e_{ij} - m_{ij}}{M_{ij} - m_{ij}} & m_{ij} < e_{ij} < M_{ij} \\ 0 & e_{ij} \leq m_{ij} \end{cases} \quad (3)$$

Treating negative indicators with lower semi-trapezoid fuzzy membership functions results in the following:

http://dx.doi.org/10.5772/intechopen.70850

$$\Phi_{(e_{ij})} = \frac{M_{ij} - e_{ij}}{M_{ij} - m_{ij}} = \begin{cases} 1 & e_{ij} \geq M_{ij} \\ \dfrac{M_{ij} - e_{ij}}{M_{ij} - m_{ij}} & m_{ij} < e_{ij} < M_{ij} \\ 0 & e_{ij} \leq M_{ij} \end{cases} \tag{4}$$

where e_{ij} is the specific attribute value of the evaluated indicator; i is the number of regions; j is the number of indicators in i region; M_{ij}, m_{ij} are, respectively, the maximum and minimum theoretical attribute values of j indicator in i region and $\Phi_{(eij)}$ is the membership level of j indicator in i region with a value between 0 and 1.

(3) Determining the integrated index potential of water resources development and the index of urban synthesized development.

With the entropy of the weight coefficient and membership values of each indicator, we can use the weighted average to individually calculate the regional integrated index potential of water resources development and the index of urban synthesized development as follows:

$$F_i = \sum_{j=1}^{m} \omega_j \times \Phi_{(e_{ij})} \tag{5}$$

where F_i is the integrated index potential of water resources development or the index of urban synthesized development of i region; ω_j is the entropy weight coefficient of j indicator relative to the highest level target, m is the number of specific indicators in the index system.

(4) Determining the water resources constraint intensity.

The correlation coefficient from the fitted equation between urbanization and total water usage will be an important basis for the measurement of water resources constraint; the smaller the correlation coefficient, the stronger the water constraint, and vice versa.

In the linear regression equation correlation coefficient, R represents the degree of coefficient between the variables X and Y. Generally, $R \in [0, 0.4]$ denotes a weak correlation; $R \in [0.4, 0.6]$ a relatively strong correlation; $R \in [0.6, 0.8]$ a strong correlation and $R \in [0.8, 1]$ an extremely strong correlation. The size of R^2, the deterministic coefficient of the regression equation, represents the degree of regression straight line fit, and $1 - R^2$ represents the deviation of the observation and fitting curve.

When the intensity of water resources constraint is 0, urbanization has a logarithmic growth curve in relation to total water consumption; with the increasing intensity of water resources constraint, urbanization increases in a less logarithmic manner in relation to total water consumption; in other words, it becomes increasingly difficult to estimate the urbanization from the logarithmic relationship to total water consumption. Therefore the percentage that cannot be estimated, $1 - R^2$, can be used as an individual measure index of the water resources constraint intensity (WRCI). This method is developed from the statistical analysis of the long-term logarithmic growth of total water consumption in relation to urbanization. If the time series is not long enough, or if the selected time period is one where the water resources usage is at low growth, zero growth or negative growth, then this model is not suitable.

In order to simplify the water resources constraint intensity (WRCI), so that it resembles the relative degree of water resources constraint on urbanization, two types of integrated indexes are combined to determine the magnitude of water resources constraint on urbanization, for example.

$$\text{WRCI} = \alpha \times (1 - F_w) + \beta \times (1 - F_u) \tag{6}$$

where F_w is the integrated index potential of water resources development; F_u is the index of urban synthesized development and α, β are the systematic influence sharing coefficients ($\alpha + \beta = 1$).

According to the fundamental principles of AHP, as far as the arid regions of northwest China are concerned, their water resources systems can be considered 'very important' to their urban systems (importance level 7), such that $\alpha = 0.875$, $\beta = 0.125$.

Based on the synthesized measurement model for water resources constraint on urbanization, the WRCI can be graded as a weak constraint WRCI ∈ [0.0, 0.3), a relatively strong constraint WRCI ∈ [0.3, 0.5), an intense constraint WRCI ∈ [0.5 0.7) or an extreme constraint WRCI ∈ [0.7, 1.0). Further, each type of WRCI can be classified as one of three grades, such as high, medium or low, for example, intensive constraint-high, intensive constraint-medium and intensive constraint-low.

4. Results and discussion

4.1. Evaluation on water resources system in Gansu section

Evaluate the water resources system in Gansu section, respectively, from the aspects of water resources series elements (rainfall, surface runoff, groundwater, total volume of water resources and water quality), water resources endowment (water production coefficient, water production modulus, agricultural water resources and per capita water resource) comparison among cities, the development and utilization of water resources (water utilization structure, water utilization profit) dynamic changes and so on [10].

4.1.1. Rainfall

Adopt trend analysis method according to the monitoring data from 75 effective meteorological stations (1956–2007), and transfer those station data into the three-dimensional perspective diagram taking property value as height, then project those points to the orthogonal surface with the map surface in two directions, fit each direction by a polynomial, and analyze the overall trend of data set. The overall rainfall in Gansu section shows the decline trend from east to west and from south to north, and the rainfall reduction between south and north is more than that between east and west.

http://dx.doi.org/10.5772/intechopen.70850

The minimum of actual measured rainfall occurred in Xihu Station of Jiuquan City in 2007 at 3.6 mm. The average rainfall remains 130.4 mm in Inland River Basin for many years, and the total rainfall is 35,214.7 million m^3, which have occupied 28.0% of the multi-year average total rainfall in the whole province; and the average rainfall remains 463.0 mm in Yellow River Basin for many years, and the total rainfall is 67,549.5 million m^3, which have occupied 53.7% of the multi-year average total rainfall in the whole province. The specific conditions of rainfall in each Basin are shown in **Table 2**.

The rainfall in each city is compared with multi-year average. Cities with an increase range over 20% include Jiuquan City, Jiayuguan City, Zhangye City, Jingchang City, Wuwei City, Lanzhou City and Baiyin City, while the increase under 10% include Dingxi City and Tianshui City.

4.1.2. Surface runoff

The water resources in Gansu section mainly and, respectively, belong to two basins of the Yellow River and Inland River. There are five water systems such as Tao River, Huang River, the main stream of Yellow River, Wei River and Jing River in the Yellow River basin; while there are three water systems such as Shiyang River, Heihe River and Shule River (including the water system of Sugan lake) in Inland River Basin. There are 36 tributaries in Yellow River Basin except for the main stream crossing through the middle part of the Province. The Basin is with large-area drainage and advantageous hydraulic conditions and 31 rivers have an annual runoff of greater than 100 million m^3. Most areas within the Basin are covered by loess, with sparse vegetation, serious water and soil loss and high sediment concentration in the river. The Inland River Basin includes three water systems like Shiyang River, Heihe River and Shule River, the basin area is 270,000 m^2. Most river sources originate from Qilian Mountain, and the northward flow and westward flow empty into the Inland Lake or disappear in desert or Gobi. Featured with short flow path, large water volume and strong current in upstream, and shallow river valley, small water volume and diverse riverbed in downstream, but the water volume is stable and the hydro power resources are abundant (**Figure 2**).

Basin	Precipitation		Multi-year average rainfall (100 million m^3)	Comparison with last year (±%)	Comparison with multi-year average (±%)	Frequency (%)
	mm	100 million m^3				
Shule River	119.0	202.28	163.189	64.6	24.0	
Heihe River	247.4	146.84	102.298	53.1	43.5	
Shiyang River	279.0	113.52	86.66	24.1	31.0	
Inland River	171.3	462.64	352.147	49.1	31.4	6.0
Yellow River	501.1	731.132	675.495	16.9	8.2	28.9
Total	310.6	1411.466	1258.306	24.9	12.2	11.8

Table 2. The rainfall conditions of each basin in Gansu section (2007).

In 2007, the natural water volume in Inland River Basin is 70.68 billion m^3, the converted runoff depth is 26.2 mm, up by 24.8% compared with multi-year average self-produced water volume. Compared with the multi-year average value, the natural water volume in Shule River increases by 26.5%, Heihe River increases by 25.2% and Shiyang River increases by 22.0%. The self-produced water volume of the Yellow River Basin is 110.87 billion m^3, and the converted runoff depth is 76.0 mm, down by 11.4% compared with multi-year average self-produced water volume. Compared with the multi-year average, the largest increase is above the Xiangtang Station of Datong River, with an increase of 57.8%. The largest reduction is above Zhangjiachuan of Jing River, with a decrease of 43.1%.

Self-produced water resources in the cities of Gansu section are compared with the average amount of years, the five cities in Hexi corridor have increased by 18–39%. The self-produced water resources of Lanzhou City, Baiyin City, Dingxi City and Tianshui City have decreased in different degrees, ranging from 20 to 40%.

4.1.3. Groundwater

The amount of groundwater resources in the Inland River Basin is 5.9706 billion m^3 (2.3667 billion m^3 in the mountainous area, 5.3314 billion m^3 in the plain area and 1.7275 billion m^3 is repeatedly calculated), which is 47.2% higher than the multi-year average of 4.057 billion m^3. Pure groundwater resources is 577.6 million m^3, increasing by 23.6% compared with the multi-year average of 467.5 million m^3.

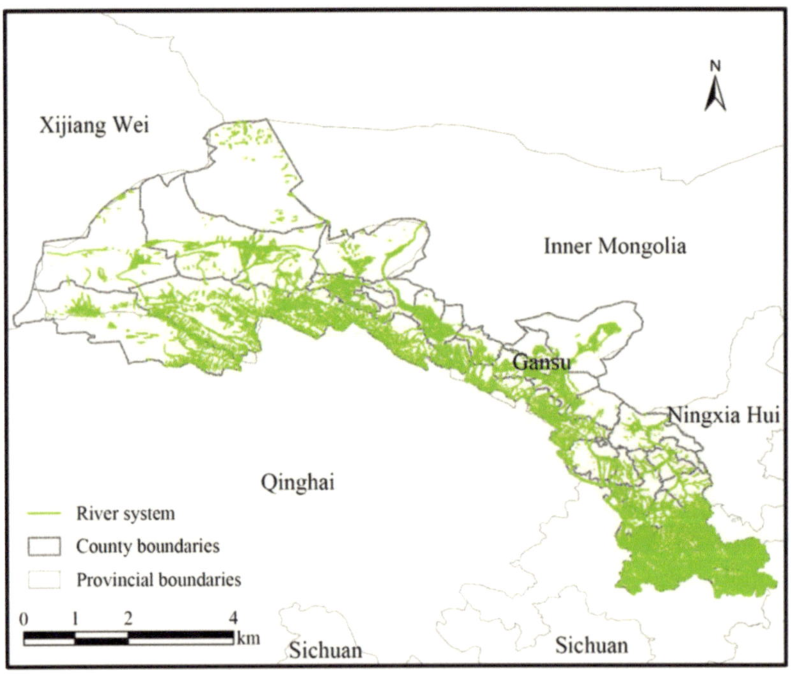

Figure 2. The spacial pattern of surface runoff in Gansu section.

The amount of groundwater resources in the Yellow River Basin is 4.5299 billion m^3 (4.4593 billion m^3 in mountainous areas, 109.8 million m^3 in plain areas and the repeatedly calculated is 39.2 million m^3), which is 0.3% higher than the multi-year average of 4.518 billion m^3. Pure groundwater resources is 387.9 million m^3, increasing by 47.7% compared with the multi-year average of 262.7 million m^3.

4.1.4. Total water resources

In 2007, the surface water resources in the Inland River Basin of Gansu was 7.068 billion m^3, the groundwater resources was 5.9706 billion m^3, the repeatedly calculated amount of the surface water and groundwater was 5.393 billion m^3 and the total amount of water resources was 7.6456 billion m^3, which was 24.7% higher than the multi-year average of 6.1291 billion m^3. Basin water production coefficient is 0.17 and water production modulus is 28,300 m^3/km^2.

The amount of surface water resources in the Yellow River Basin is 11.087 billion m^3, the amount of groundwater resources is 4.5299 billion m^3, the repeatedly calculated amount of the surface water and groundwater is 4.142 billion m^3, and the total amount of water resources is 11.4749 billion m^3, which is 22.6% higher than the multi-year average of 10.0364 billion m^3. Basin water production coefficient is 0.36, water production modulus is 201,900 m^3/km^2.

Affected by both climate change and regional development on the use of water resources, the annual runoff of the basins is generally showing a downward trend, and some basins decreased significantly, indicating that during the process of urbanization, the conflict between urban water supply and demand will become more and more prominent, and the pressure of water resources will be mounting.

4.1.5. Water quality

In 2007, 68.1% of the cross sections of rivers in Gansu section reached the water quality standard. The water in Shiyou River of Yumen City; Shandan River of Zhangye City was increasingly polluted. The water quality of Lanzhou section in the Yellow River was Category III; and in the Baiyin section, the chemical oxygen demand and ammonia nitrogen concentration in three cross sections had been slightly decreased compared with last year, and the water quality was raised from Category III to Category II. In Tianshui section of the Weihe River, the chemical oxygen demand and ammonia nitrogen concentration had been decreased in other three cross sections excepted the Hualin cross section with the water quality being improved compared with last year. The comprehensive indexes for pollution in each cross section in Zhangye section of the Heihe River remained the level last year with excellent water quality. The concentration in such indexes as permanganate index, chemical oxygen demand, biochemical oxygen demand, ammonia nitrogen, and total phosphorus in Shiyang River had been decreased. However, the total water quality pollution in the reach was relatively serious remaining the Category Bad V. The biochemical oxygen demand concentration in Hongyashan Reservoir was increased with the water quality pollution being slightly aggravated.

In the water quality of drinking water source, the water quality of the centralized drinking water sources in nine cities was excellent totally. Except the fecal coliform which exceeded the standard in January, June, August and September in Baiyin City, the drinking water quality in each monitoring month in other eight cities had reached the standard. Through protecting the drinking water sources and gradually reducing the influence of human activities on the water sources, the water environment situations of the drinking water sources in each city have been improved recently to some extent.

4.1.6. Comparison for water resource endowment of cities

Influenced by multiple factors, the imbalance for spatial distribution of regional water resource had been studied, which had caused the obvious endowment difference in water resources in nine prefecture-level cities (**Table 3**).

The shortage of water resource was relative to the water resource demands in the region. In view of the total water source quantity hardly reflecting the abundance and shortage of water resource in a region, it might be considered to compare such factors as population, agricultural acreage and effective irrigation area and the water resource endowment in the city.

In nine cities, due to limited water resource in Dingxi City and many cultivated lands, the agricultural allocated water resource quantity was the lowest (only 484.24 m^3/mu). Due to limited cultivated land resource in Jiayuguan City, the agricultural allocated water resource quantity was the highest reaching 1246.93 m^3/mu. The difference exceeded 2.5 times that the discrepancy in water resource conditions of agricultural development in the research region was obvious. In addition, respectively, compared with the whole Gansu Province (11,240.2 m^3/mu), the western Longhai-Lanxin economic zone (10,094.0 m^3/mu) and the whole country (12,386.4 m^3/mu), the total agricultural water condition in the research region was poor (**Figure 3**).

The difference in water resource quantity per capita in nine cities was very large. Lanzhou City has less water and large population size resulting in the low water resource per capita, while

City	Annual rainfall (mm)	Surface water (100 million m^3)	Underground water (100 million m^3)	Repeated calculation amount	Total amount of water resource	Proportion (%)	Runoff coefficient	Runoff odulus (10,000 m^3/ km^3)
Lanzhou	55.396	1.491	1.035	0.359	2.167	0.80	0.04	1.60
Baiyin	68.067	1.122	1.627	0.242	2.507	0.93	0.04	1.25
Tianshui	83.140	13.240	5.571	5.237	13.574	5.05	0.16	9.48
Dingxi	96.530	8.344	3.980	3.484	8.840	3.29	0.09	4.51
Wuwei	89.252	13.430	10.039	8.331	15.138	5.63	0.17	4.55
Zhangye	138.93	36.749	26.416	24.491	38.674	14.38	0.28	9.45
Jianchang	20.785	0.614	2.244	1.969	0.899	0.33	0.04	1.18
Jiuquan	229.47	26.048	22.799	21.049	27.778	10.33	0.12	1.45
Jiayuguan	2.167	0.013	0.430	0.360	0.083	0.03	0.4	0.64

Table 3. Comparison for water resources in various cities of Gansu section (2007).

Jiayuguan was extremely short of water resulting in the worst situation per capita. Similarly, compared with the whole Gansu Province (1150 m^3/person), the western Longhai-Lanxin economic zone (1660 m^3/person) and the whole country (2400 m^3/person), for the water resource per capita in nine cities, only two cities had better situations and the rest was poor (**Figure 4**).

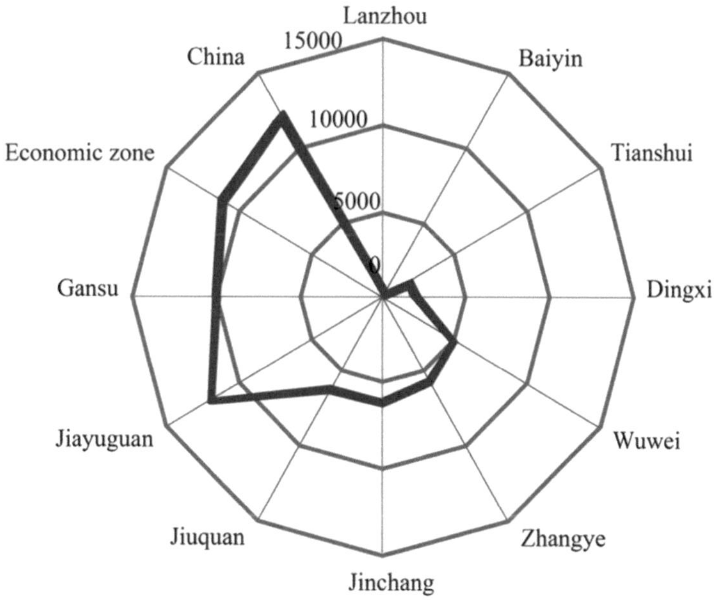

Figure 3. The agricultural allocated water resources.

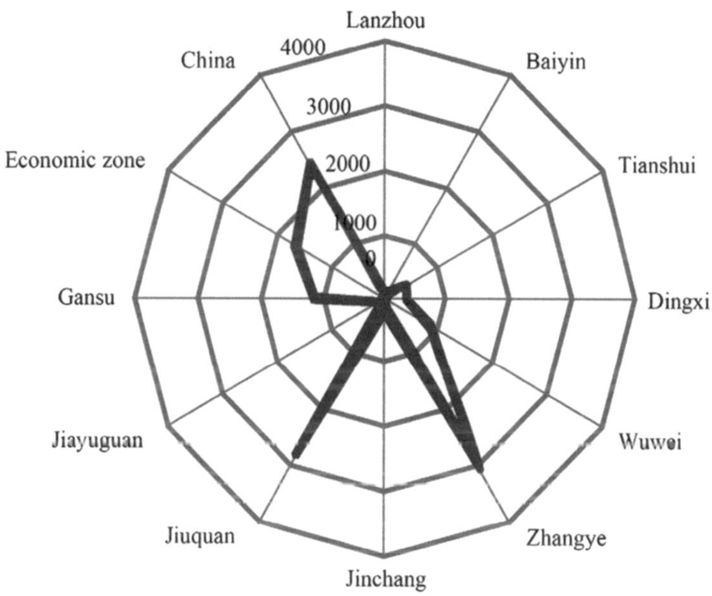

Figure 4. The water resources per capita.

4.1.7. Water resource development and utilization

The ways to use the water resource generally contained three aspects, namely agricultural water, industrial water together with public water in cities and domestic water of residents. Due to different development scales, development degrees and water demands among nine cities in Gansu section, corresponding difference in the water resource consumption was obvious. Besides, the changes for three types of water utilization in several key phases were very large with all types of water utilization presenting the increase trend as a whole. In the water utilization structure, except more industrial water used in Lanzhou City and Jiayuguan City almost accounting for a half, other cities mainly utilized the water in agriculture, especially in Wuwei, Zhangye and Jiuquan located in the Inland River Basin, the proportion of agricultural water exceeded 90%, which were the typical agricultural development-oriented regions. Besides, there were large differences in the water transfer directions and ranges in each city related to the specific situations of industrial development in each city.

The water resource efficiency is an important index reflecting the development and utilization of water resource and an index reflecting the development and utilization potential of water resource in a region. Two indexes, namely agricultural water consumption per 10,000 yuan output value and industrial water consumption per 10,000 yuan output value were adopted here to compare the water resource utilization efficiency in each city in Gansu section for many years.

As a whole, the agricultural and industrial water utilization efficiency in most cities were increasing with large increased range in the industrial water utilization efficiency. There were abnormalities in some cities: in Lanzhou City and Jinchang City, the agricultural water consumption per 10,000 yuan output value was lower in other years than that in 1991 relating to small irrigation area at that time; and in Wuwei City, the lower industrial water consumption per 10,000 yuan output value in 1991 was caused by seriously insufficient industrial development at that time with relatively less water consumption.

In 2007, the agricultural and industrial water consumption per 10,000 yuan output value in Gansu Province were 3414.01 m^3 and 104.07 m^3, respectively. Accordingly, in western Longhai-Lanxin economic zone, the corresponding values were, respectively, 5322.45 m^3 and 141.62 m^3; and in the whole country, the values were, respectively, 1301.25 m^3 and 121.48 m^3. Compared with them, the situations in each city in Gansu section were good or bad (**Figure 5**) reflecting the obvious relative difference in water-saving level and efficiency among them.

4.1.8. Comprehensive judgment for abundance degree of available water resources

The abundance degree of available water resource quantity was comprehensively judged from such information as spatial and temporal distribution and development and utilization degree of water resource and water composition in 42 counties in Gansu section. In order to more directly show different abundance degree among counties, **Figure 6** could be applied for analysis that in 42 counties in Gansu section, 25 of them had the situations in and below the medium almost accounting for 60% reflecting relatively insufficient water resource in most of regions with poor water utilization conditions. In addition, the counties with more available water resource quantity were mainly located at the south of Gansu section in space, while the

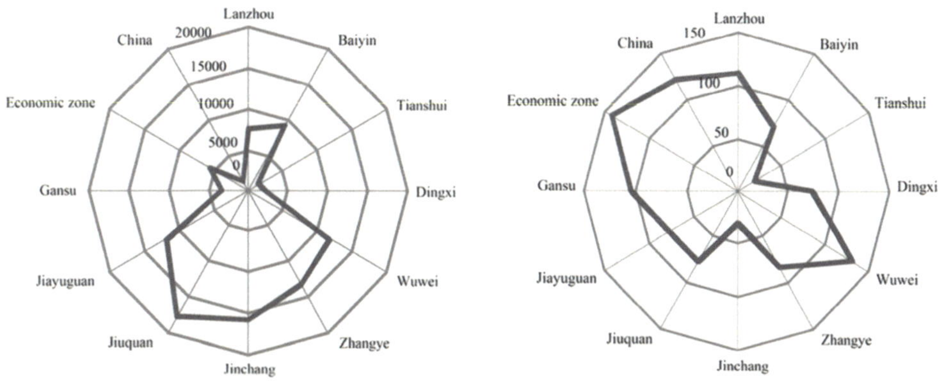

Figure 5. Comparison for water consumption per 10,000 yuan output value (left: agriculture; right: industry).

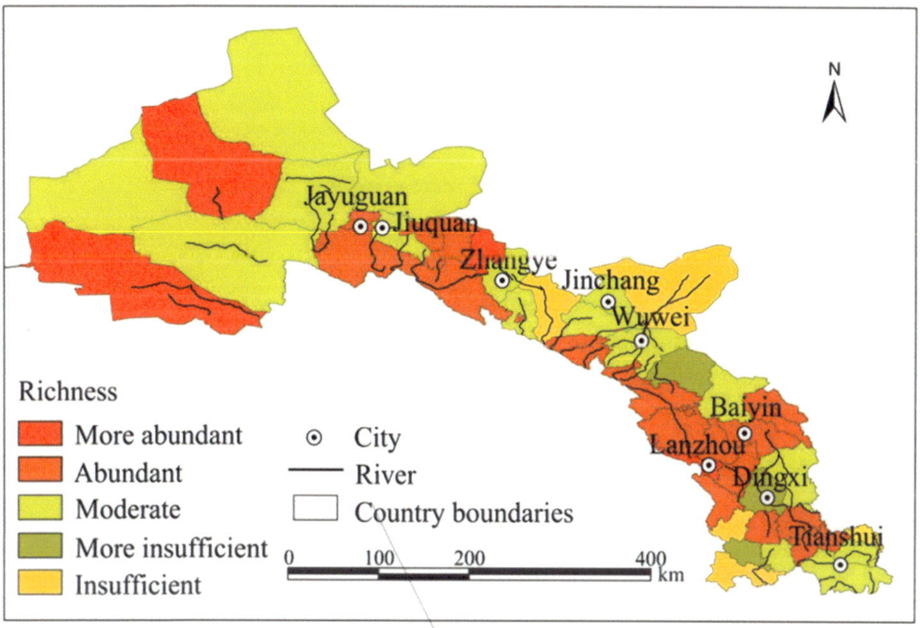

Figure 6. Comprehensive judgment for richness of available water resources in 42 counties of Gansu section.

economic center of Gansu section was mainly centered on the middle part along the axial zone of Lanzhou-Lianyungang Railway. Therefore, the water resource and economic development in Gansu section were poorly matched spatially.

4.2. Evaluation on urbanization system

The urbanization had diversified meanings, mainly including four aspects, namely population urbanization, economic urbanization, social urbanization, and spatial urbanization. The population urbanization was the foundation. The population urbanization referred to the dynamic process of rural population being converted and centered to the city and the proportion of urban population being gradually improved, mainly including three ways, namely natural increase in urban population; a batch of rural population pouring into the city and rural population locally converted into the population in urban life style through the social and economic development. Due to different types of society, economy, culture and ecology in all cities, the progress, characteristics, ways and social consequences of population urbanization were definitely and largely different [11].

4.2.1. *Dynamic changes in population urbanization development in all cities*

Due to diversified historical relics, local and natural resource conditions, economic and social development basis and investment environment in all cities, differences existed in the urbanization in each city. Compared each other in several selected years, the differences in the population urbanization could be seen (**Figure 7**).

Several cities with higher urban level and faster increased range were successively: Jiayuguan, Lanzhou and Jiuquan, and other six cities were approximately same without large difference reflecting the relatively stable boost of population urbanization in Gansu section as a whole with few prominent regions.

4.2.2. *Dynamic changes in population urbanization development in all counties*

In different periods, the difference of the population urbanization progress in counties was relatively obvious with rich contents. After respective comparison of urbanization level in each county in 1995 and the average growth rate of urbanization in previous 5 years with those in 2007 accordingly, the changes and differences in the development trend for urbanization progress could be judged (**Figure 8**).

The urbanization level in most of counties in Gansu section was lower than the average level in 1995. There were only 15 counties higher than the average level indicating the obvious

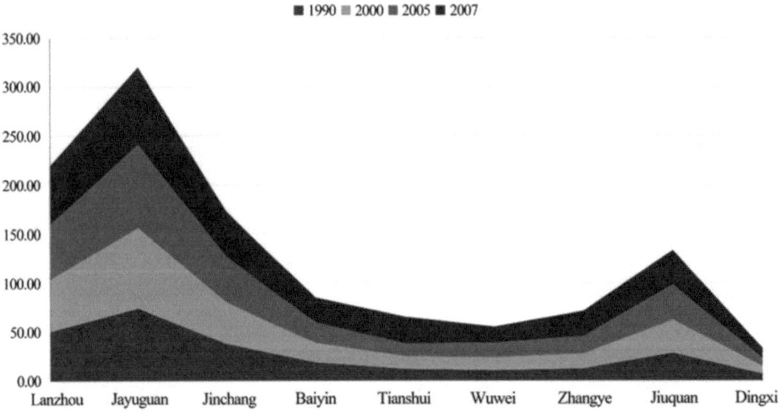

Figure 7. Cumulative curve for population urbanization expansion in nine cities in Gansu section.

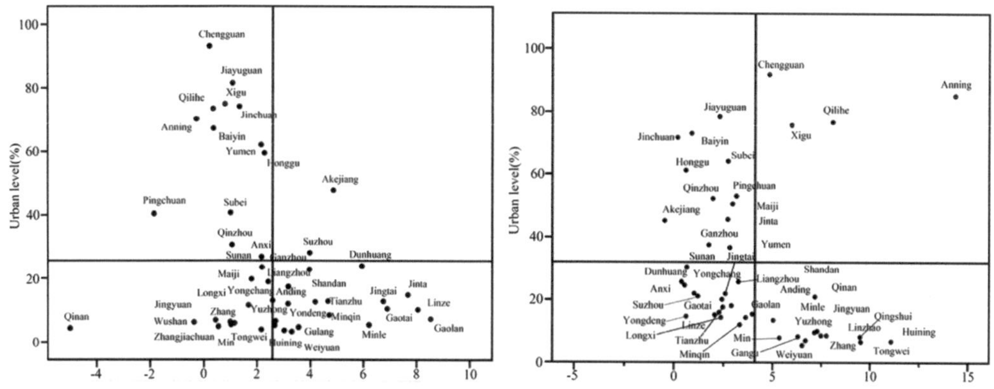

Figure 8. Difference of level and average growth rate of population urbanization in 42 counties.

polarization; and there were many counties with the growth speed higher than the average level reflecting good development trend as a whole with certain potential growing group existed. Only Kazak Autonomous County of Aksay and Suzhou District had high level and high growth speed with less influence. Until 2007, the urban level and urban growth of all counties and districts were distributed dispersedly reflecting the shrinkage of group difference; and the number of counties or districts having higher level and higher growth speed were increased to four, which were the areas under administration of Lanzhou city, the capital city, reflecting strong centralization of the central city.

4.2.3. Comparison of population urbanization

Place the population urbanization of Gansu section, respectively, in different background (Gansu Province, the economic zone, the Northwest China, the western China and China) for comparison, to judge its process speed and determine the opportunities and challenges it was faced with (**Figure 9**).

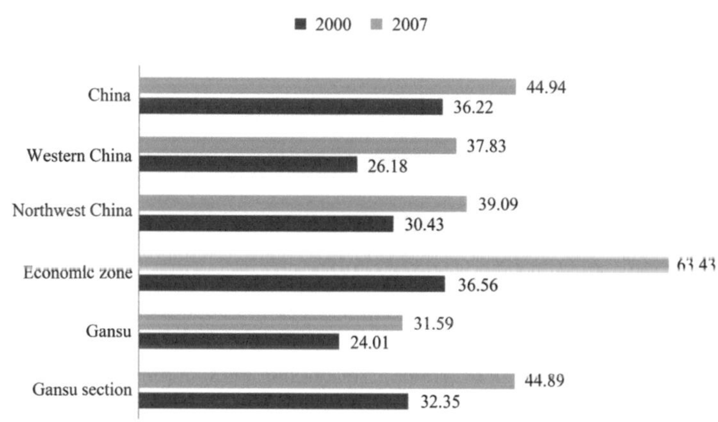

Figure 9. Comparison of population urbanization.

As s a prior-developed area in the northwest China, the urbanization level of Gansu section in 2007 (44.89%) was obviously higher than the level of the whole province in the synchronized period (31.59%), and it was also higher than the average levels in Northwest China and western China (39.09%, 37.83%), it was very close to the national average level (44.94%), but it was still lower than the level of western Longhai-Lanxin economic zone (63.43%), which reflected the developing depression situation of the Gansu section as economic zone was continuously existing. From 2000 to 2007, the urbanization level of Gansu section has increased by 1.79 point each year, and it was, respectively, higher than that of the whole province (1.08), the Northwest China (1.24), the western China (1.66) and the National (1.25), which reflected the Gansu section belongs to the area with good developing environment compared with large-area background, but its increasing speed of urbanization rate was lower than that of the economic zone (3.84), which also reflected that the competition situation of Gansu section in the whole economic zone was relatively poor and the development was relatively slow. For the Gansu section, which promotes economic recovery by urbanization, the high-speed development for the urbanization is essential in the future and the restraints of resource like water resources arising from it will be increasingly prominent.

4.3. Temporal-spatial variation of water resource restraints

4.3.1. Results and analysis of F_w and F_u

Based on the above formula (5), we can calculate the synthesized index potential for water resources development F_w for the Gansu section and for each city within it. In view of the physical significance of F_w, a value of 1 for F_w indicates that the water resource system was broadly undeveloped, and that all the water resources were left for the eco-environmental system; a value of 0 for F_w indicates that 100% of the water resources were developed and used for the socioeconomic system, and no water resources were left for the eco-environmental system. F_w shows a positive development, particularly since 2002, in the Gansu section, reflecting that after years of efforts such as water conservation and appropriate allocation, etc., the socio-economic system started returning water to the eco-environmental system, resulting in nearly 30% (the average value of F_w was 0.2972) of water resources being left for the eco-environmental system. There are, however, F_w differences among the nine cities. Both Lanzhou and Jiuquan had relatively smaller F_w values for different reasons. Lanzhou city's value is the result of environmentally friendly urban development, whereas Jiuquan city's value was caused by a low level of water development. The remaining seven cities had relatively equivalent and larger values of F_w. Wuwei city, one of the Inland River (Shiyang River) cities, has been increasing its proportion of water left for the eco-environmental system through large national and local efforts on the reasonable use of water. But compared to the international alert level of 40% water resources development (i.e., 60% of water remains for the ecosystem), all nine cities of the Gansu section should return more water from the socioeconomic system to the ecosystem [7].

Similarly we can calculate the synthesized index of urban development F_u for the Gansu section and for each city within it. There has been a continuous increase of F_u in the Gansu section and in each city within it for the last 20 years. The F_u of the Gansu section has risen from slightly more than 0.3 in 1989 to nearly 0.5 in 2007, an average annual increase of more than 0.01, reflecting rapid urban development in the Gansu section. Among the nine cities, the

http://dx.doi.org/10.5772/intechopen.70850

F_u of four cities: Lanzhou, Jiayuguan, Jingchang and Tianshui: is relatively larger than that of the other five cities: Jiuquan, Zhangye,Wuwei, Baiyin and Dingxi, reflecting unbalanced urban development in the Gansu section.

Furthermore, comparison between F_w and F_u shows development changes and coupling trends over time. In general, in the last 20 years, after various efforts, there have been sustained increases in the level of Gansu section's urban development, and a gradual drop in the development level of water resources (moderate increases of corresponding F_w). Especially in the twenty-first century, less than 70% of the water resources have been developed, although this level is still much higher than the international alert level of 40%. At the same time it can be seen that the rise in the Gansu section's F_w is clearly smaller than that of its F_u, so that the relative constraint of water resources on urbanization has not declined, and water resources management will become increasingly difficult if this trend is not reversed [7].

4.3.2. *Comprehensive measurement of water resources constraint on urbanization*

Using formula (6) we can determine the individual intensities of water resources constraint on urbanization for the Gansu section and its nine cities (**Table 4**).

Year	Lanzhou	Jiayuguan	Jinchang	Baiyin	Tianshui	Wuwei	Zhangye	Jiuquan	Dingxi	Gansu section
1989	0.6463	0.7095	0.7122	0.7628	0.6616	0.7876	0.7961	0.7840	0.7174	0.7309
1990	0.6408	0.7219	0.7229	0.7792	0.7148	0.7975	0.8078	0.7193	0.7342	0.7376
1991	0.6334	0.7235	0.7033	0.7608	0.6709	0.7806	0.7926	0.7266	0.7167	0.7231
1992	0.6260	0.6811	0.7076	0.7586	0.6668	0.7445	0.7700	0.6940	0.7299	0.7087
1993	0.6178	0.7312	0.6975	0.7521	0.6887	0.7593	0.7618	0.6915	0.7091	0.7121
1994	0.6107	0.7186	0.6947	0.7481	0.7076	0.7621	0.7710	0.7259	0.7453	0.7204
1995	0.6023	0.7154	0.6982	0.7472	0.7227	0.7619	0.7661	0.6943	0.7424	0.7167
1996	0.5948	0.7094	0.6996	0.7493	0.7232	0.7528	0.7568	0.7021	0.7484	0.7151
1997	0.5855	0.7114	0.6918	0.7441	0.6954	0.7402	0.7475	0.7171	0.7395	0.7080
1998	0.5886	0.7177	0.6885	0.7402	0.6978	0.7457	0.7440	0.6690	0.7401	0.7035
1999	0.5896	0.7141	0.6998	0.7513	0.6889	0.7363	0.7557	0.6755	0.7321	0.7048
2000	0.5855	0.6975	0.6834	0.7527	0.6686	0.7281	0.7393	0.6664	0.7296	0.6946
2001	0.5822	0.7073	0.6836	0.7457	0.6990	0.7149	0.7375	0.6429	0.7416	0.6949
2002	0.5723	0.6278	0.6905	0.7385	0.6084	0.6846	0.7262	0.6722	0.6909	0.6679
2003	0.5980	0.5829	0.6940	0.7484	0.6835	0.6565	0.7280	0.6813	0.7264	0.6777
2004	0.5040	0.6810	0.6652	0.7509	0.6155	0.6408	0.6895	0.6274	0.6969	0.6523
2005	0.5220	0.6733	0.6207	0.7353	0.6454	0.6566	0.7097	0.6504	0.7130	0.6585
2006	0.4919	0.6696	0.6169	0.7193	0.6204	0.6457	0.6963	0.6144	0.7029	0.6419
2007	0.4789	0.6795	0.6121	0.7231	0.6469	0.6674	0.7034	0.6344	0.6967	0.6492
Average	0.5827	0.6933	0.6833	0.7478	0.6751	0.7244	0.7473	0.6836	0.7238	0.6957

Table 4. Gansu section and individual cities' 1989–2007 annual variations of WRCI.

From 1989 to 2007 the WRCIs both for the Gansu section and for each individual city have been decreasing, albeit with much fluctuation. Since 2000, particularly, most cities had accelerating decreases in WRCI, with Lanzhou and Jingchang being the most obvious, thanks to a rational allocation of water resources, the amount of national and local investments in water-saving techniques and cyclic utilization of waste water, better water resource management and water-saving awareness, etc. There are, however, obvious differences among the WRCIs of the nine cities of the Gansu section, reflecting variations in the WRCIs. Specifically, the situations in these nine cities over this period have been: Lanzhou changed from intensive constraint to relatively strong constraint, with a significant decline in its WRCI; five cities changed from extreme constraint to intensive constraint, including Wuwei and Jiuquan with significant declines in their WRCI, Jiayuguan and Jinchang with no dramatic declines and Dingxi with a fluctuating decline; three cities remained at the same level of water resource constraint, including Zhangye and Baiyin with extreme constraint, and Tianshui with intensive constraint [7].

4.3.3. *Spatial differences among types of water resources constraint on urbanization*

Based on ArcGIS and the 20-year average of WRCI for each city, the spatial differences among the types of water resources constraint in the nine cities in the Gansu section are shown in **Figure 10**. In particular, Lanzhou, the capital city of Gansu province, belongs to the relatively strong constraint-high type, four cities including Jiuquan, Jiayuguan, Jinchang and Tianshui, belong to the intensive constraint-high type, and the remaining four cities, including Zhangye, Wuwei, Baiyin and Dxingxi, belong to the extreme constraint-low type. These

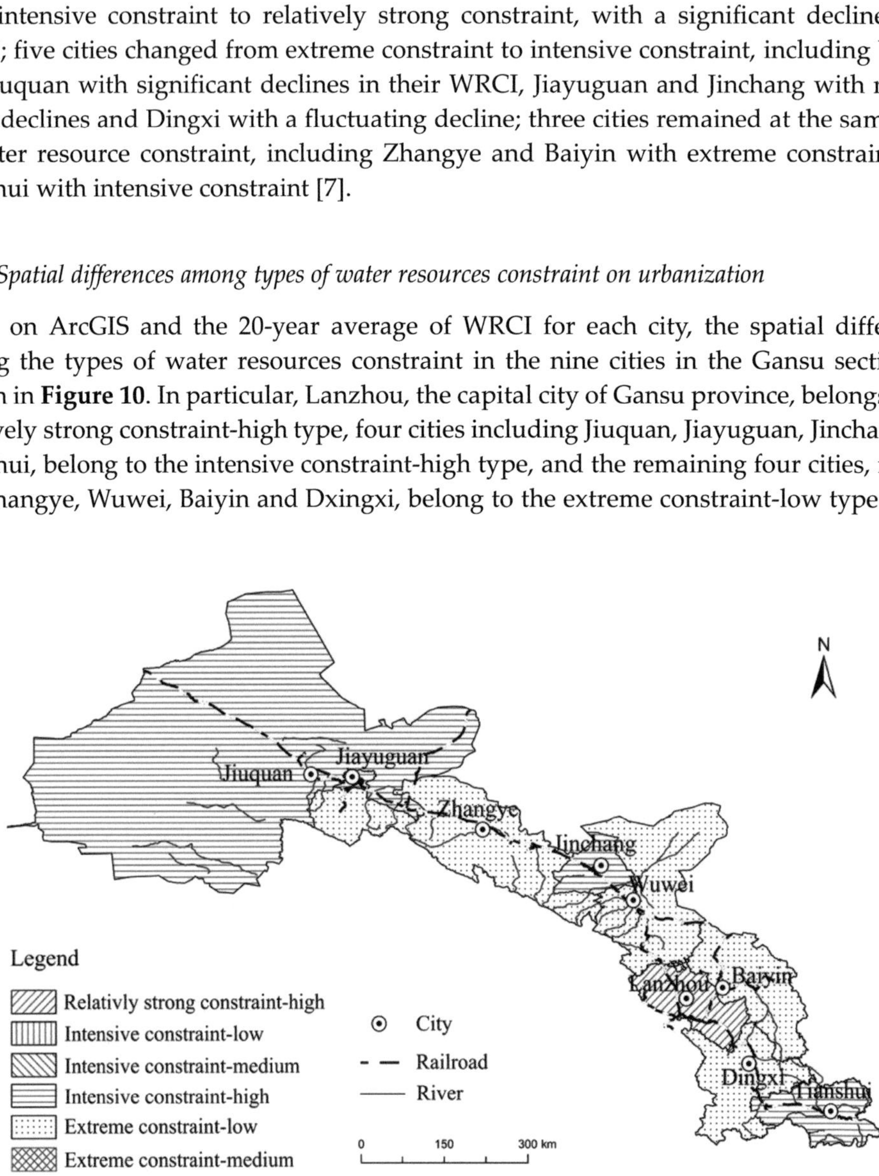

Figure 10. Spatial differences among types of water resource constraints on urbanization in cities.

results show that the Gansu section will continue to be affected by water resources constraint on urbanization and local economic development for some time to come [7].

Improving the urbanization level is the important way to reduce the strength of water resource restraints, but improving urbanization itself will not lead to the increasing strength of water resource restraints, in the certain stage, the reason why the strength of water resource restraints increases with the development of urbanization is that deterioration of water resource health conditions exceeds the improvement of urbanization system health conditions, which eventually lead to the overall deterioration of composite system health conditions.

5. Conclusion

The rainfall and surface runoff, in Gansu Section of western Longhai-Lanxin Economic Zone, showed the pattern of high in east, low in west, abundant in south and few in north, all cities were significant different in water resource endowment; for the water utilization structure, except more water for industrial utilization in Lanzhou City, Jiayuguan City, which has occupied almost half of all, the other cities mainly utilized water for agriculture, especially in the Wuwei, Zhangye and Jiuquan City of the Inland Basin, and all of their proportions of agricultural water utilization were over 90%, and they were the areas which typically oriented by agricultural development. Besides, the transferring directions and transferring amplitude of water utilization in all cities had great difference, which was related to the specific conditions of the industry development of all cities; for the water utilization benefits, although all kinds of benefits were increasing, but compared with the economic zone and the whole nation, it did not own obvious advantages and there was still difference in the internal.

The comprehensive development level of utilization was the highest in Jinchang and lowest in Dingxi, and Lanzhou ranked in the third place, and this was caused by the composition differences of respective utilization connotation. In the aspect of population urbanization of Gansu section, among nine cities, the faster increasing amplitude only existed in Jiayuguan, Lanzhou, Jiuquan and Jinchang, which reflected the whole urbanization promotion was slow and there was less significant areas; the polarization of urbanization for 42 cities was obvious, and the increasing speed higher than the average level was rather more, and this reflected that the whole development conditions were rather good and there was certain increasing potential group. Compared with the Northwest China and the whole nation, the urbanization development in Gansu section owned certain advantages, but compared with the economic zone, both of its urbanization level and increasing speed were rather low, which reflected the developing depression situation of the Gansu section as the economic zone continuously existed.

For all nine cities, the water resource strongly restrained their urbanization, to be specific, Wuwei, Zhangye, Baiyin and Dingxi were extremely strong restrained type, while other cities were strong restrains type, and Lanzhou was relatively the lowest in water resource restraints. From 1989 to 2007, the water resource restraints strength in Gansu section and all cities fluctuated certainly but the fluctuation was rather small, but since 2000, there was a dropping trend of water resource restraint strength in most cities and it tended to speed up, and it was especially obvious in Lanzhou and Jinchang City. On the multi-year average, excepted that Lanzhou was the strong type of water resource restraints (high) and Tianshui was the strong type of water resource restraints (high), the water resource restraints type of other seven cities was extremely strong (low), and this reflected Gansu section, in its overall socioeconomic development, has born high pressure of water resource, the task to coordinate water resource restraints strength was rather difficult.

Although with the development, the water resource to restrain the urbanization development of most cities has slowed down in a certain extent, in a very long time, the water resource still will be the strong limiting factors for the urbanization development in this area and it needs to be coped with reasonably. The mitigation approach to water restraints includes: build effective, water-saving, pollution-preventing and contemporary agricultural system, effective, water-saving, pollution-preventing industrial and mineral industry, effective, water-saving, pollution-preventing urban-rural system and reasonably allocate water resource.

The relationship between the urbanization and water resource is interacting and cross-coupling. On one hand, the urbanization has stress to water resources through increasing population, economic development, water resource consumption and regional expansion; on the other hand, water resources restrains the development of the city in the way of industry locations, population migration, water resources restraints, and fighting for water resource development fund and policy intervention of water resource development. Water resources system and the urbanization are two aspects of one relationship, in the past, most Chinese scholars placed the emphasis of their research on the stress of cities' development to water resource, and this research, rarely from the view of water resources, kept the idea of water resource restraints throughout the research to quantificationally evaluate the spatial-temporal change of urban process and strength from macro scales (provincial level), and this is not only the beneficial complement for that the research of water resource bearing capacity and water resource pressure is inadequate to cover the interaction between water resource system and socioeconomic system, but also beneficial to expanding the economics research of macro-increasing economics to provide decisive basis to realize the coordination and sustainable development for the water-ecology-society system.

Two large systems: water resource and urbanization built in this research have involved over 2000 data, the data amount is large, the individual data source is inconsistent, and in addition, the availability of individual index and data is poor, so the data for part of the years is obtained by adopting interpolation method and tendency method, and the objectivity is insufficient, all above need to be further improved.

With the promotion of the Silk Road, the Gansu section of western Longhai-Lanxin economic zone will also enter into a new and fast-speed development period, and its water resource pressure and restraining issues during the development will be more prominent. Estimating the future to coordinate the present, and research emphasis in the next step will comprehensively integrate all kinds of qualitative and quantitative analysis method and simulate the restraint mechanism of water resource in the future to the urbanization and forewarn the sustainable development model in different circumstances of water resources.

Acknowledgements

Parts of this chapter are reproduced from the author's EES journal article [7]. The work described in this chapter was supported by the State Key Program of National Natural Science of China (Grant No. 41530745). I would like to express my gratitude to Professor WANG Naiang from Lanzhou University, for his constructive suggestions, and give my thanks to Dr. Yu Tengfei from Northwest Institute of Eco-Environment and Resources, CAS, for his technical support.

Author details

Gao Xiang

Address all correspondence to: xgao@lzu.edu.cn

College of Earth and Environmental Sciences, Lanzhou University, Lanzhou City, China

References

[1] Biswas AK. Water for sustainable development in the 21st century: A global perspective. International Journal of Water Resources Development. 1991;**4**:219-224

[2] McMichael AJ, Haines A, Slooff R, Kovats S, editors. Climate Change and Human Heath. Geneva: World Health Organization; 1996. p. 297

[3] Changming L, Xi-wu H. Strategies on Water Problems in 21st Century in China. Beijing: Science Press; 1996. p. 77-81

[4] World Water Council. World Water Vision, London: Earthscan Publications Ltd; 2000. p. 46-70

[5] Wen-lai J. Study on water resources safety strategy for China in the 21st century. Advance in Water Resources. 2001;**12**:66-71

[6] Chuang-lin F, Xiao-lei Z. The development orientation and the industrial division of the node cities along the western long Hai-Lan Xin economic zone. Geographical Research. 2003;**22**:455-464

[7] Gao X, Tengfei Y. The temporal-spatial variation of water resources constraint on urbanization in the northwestern China: Examples from the Gansu section of west Longhai–Lanxin economic zone. Environment and Earth Science. 2014;**71**(9):4029-4037

[8] Bao C, Fang CL. Temporal and spatial variation of water resources constraint on urbanization in arid area. Acta Geographica Sinica. 2008;**163**:1140-1150

[9] Bao C, Fang CL. Scenario warning and analysis of water resources constraint intensity on urbanization in arid area. Journal of Natural Resources. 2009;**9**:1509-1519

[10] Chuang-lin F, Chao B, Biao Q. Urbanization Process and Its Eco-environmental Effects. Beijing: Science Press; 2008. p. 97-124

[11] Chuang-lin F. Urbanization Process of China and Resource-Environment Report. Beijing: Science Press; 2008. p. 35-58

Chapter 8

Sentinel-1 Satellite Data as a Tool for Monitoring Inundation Areas near Urban Areas in the Mexican Tropical Wet

Alejandra A. López-Caloca,
Felipe Omar Tapia-Silva and Guadalupe Rivera

Additional information is available at the end of the chapter

http://dx.doi.org/10.5772/intechopen.71395

Abstract

This work shows advances in the field of water body monitoring with radar images. Particularly, a monitoring procedure is developed to define the extension and frequency of inundation for continental waters of the Grijalva-Usumacinta basin, in the state of Tabasco, Mexico. This is a region located in the Mexican tropical wet and under its meteorological conditions, radar technology can be used to characterize monthly inundation frequency. The identification of water bodies were obtained by processing images at a monthly intervals captured by Sentinel-1A during 2015 having kappa indices and overall accuracy higher than 0.9. The chapter describes the seasonal variability of these water bodies, and at the same time, the relationship with human settlements located in their neighborhood. To do this, a proximity analysis was carried out to emphasize the importance of spatial-temporal studies of superficial water bodies, linked to an urban and a rural area. This information is useful to investigate changes in the ecosystem, as well as risks to human settlements, and as a contribution for a comprehensive management of hydric resources.

Keywords: Sentinel 1A, water body, urban areas, Tabasco, Mexico

1. Introduction

The location and evolution of cities are defined by the geographic spaces where they are established, so that they generate well-being for their inhabitants through services such as: drinking water, food, power, transportation, recreation, among other services [1]. The dynamics of urban growth over time modify the spatial form and organization of the urban space, as well as of the surrounding natural and rural areas. Different authors agree that urbanization presents problems pertaining to water demand, land-and land cover changes [2, 3], irregular human settlements, pollution, and the production of wastewater [4]. Thus, rapid urbanization on the land's surface leads to the ecological instability of natural resources, compromising and overwhelming their capacity for providing environmental services. The consequences of exceeding the capacity of natural resources are the degradation, fragility, and loss of biodiversity, due to an increase in the demand for natural resources. At the same time, climate change problems arise.

Historically, the foundation and development of urban spaces has had a direct relationship to the closeness of natural sources of drinking water, since many important cities are located at the edges of great lakes, or along the coast and rivers. To date, research on the impact of urbanization on lakes and rivers, includes works documented at the urban ecology, urban catchments, and landscape ecology levels [5–7]. Several cities have had an impact on hydric resources, modifying the course of rivers and their connection with the stream that feeds them. To increase the urban land area, the expansion has led to the implementation of drainage processes, drying out, and the losing of surface water bodies. The elimination of vegetation changes the permeability of soils, impermeable soils increase runoff and the accumulation of water increases when a suitable drainage network is not available.

To understand the conservation and loss of water body surfaces, Dunne [8] proposed an evaluation of how cities create changes in land-use and their effects on hydrography. Yue [9] suggests investigating urban rivers by considering fundamental themes such as the characterization of the hydrological pattern and function. In the U.S. National Context, the study by Steele and Heffernan [10], related the patterns and characteristics pertaining to shape, form and connectivity of surface water and urban spaces in 100 U.S. cities. Based on their location and attributes, water bodies on urban land covers tend to reduce their size, and in the urban space there are also fewer water bodies connected to streams and rivers, compared to non-urbanized lands. Small water bodies are mostly affected by urbanization.

The study of seasonal patterns of water bodies is an indicator of the natural behavior of water and helps society in decision-making with respect to water management, providing more information on when and how the natural protected areas must be established and for building basic infrastructure (dams, dikes, and bridges). This will allow for a decrease in possible losses due to flooding, in terms of life, property, and businesses. An example of one study regarding the importance of water body monitoring in Tabasco, México is found in the work by Rodríguez [11], who make an inventory of continental lagoons in Tabasco, in the year 2002. The study shows the importance of the information on surface water areas and their water body morphological analysis, applied to a model of fishing efficiency.

In view of this situation, we need to explore more sustainable alternatives in terms of the relationship between surface waters and metropolitan areas [12], with a focus on Integrated Water Resources Management (IWRM). This is defined as *"a process that promotes the coordinated development and management of hydric, land and other resources, in order to achieve maximum economic and social well-being in an equitable manner, without compromising the sustainability of vital ecosystems"* GWP [13]. Savenije and Van der Zaag [14] described four key factors in IWRM, which are: the water resources, the water users, the spatial scale, and the temporal scale. The first two refer to the natural behavior of water, to human aspects and requirements, while the third key factor points to the spatial geographic distribution and the hydrology of water bodies. Some regions and countries such as Africa [15], Holland, and Great Britain have broad knowledge and experience with geographic space, water management, and logistics, to face problems pertaining to excess water due to precipitation or drought. The last key factor refers to the behavior and temporal variability of water, as well as to its demand, a factor affected by drought or flooding. The consequences of surface water redistribution at spatial and temporal scales, including the landscape level, are not well known. Steele and Heffernan [10] found that cities gain or lose water, so that they become different from their surrounding natural landscapes, in terms of the spatial configuration. There is a lack of studies to investigating the spatial and temporal variability of water bodies and their effects on urban areas.

The objective of this investigation is to contribute to the analysis of this process, based on the study of the inter-relationships between water dynamics in lakes and rivers and urban space in a part of the Mexican tropical wet. The chapter analyzes the dynamics of surface water bodies in the low plains of South-Eastern Mexico. The urbanization process that happened in the low lands of the region and in the flatlands, far from being a planned growth, obeyed economic needs and the need for land for housing. Because of their geographic location, the dynamics of these territories involve frequent flooding area changes in relatively short times because of the high volume observed during the rainy months.

In this sense, a monthly spatial-temporal analysis is set forth, of water bodies that are linked to rural and urban zones of Villahermosa, in the state of Tabasco. The research strategy uses satellite images, since these are powerful tools to identify the geographic features, such as places and natural aspects. These new satellite technologies facilitate processing and extraction of information for consistent, multi-temporal measurements of water body surfaces. Through the processing of Sentinel 1A radar images, the spatial and temporal variability of the water bodies' footprint was defined. The radar data allow us to have a closer look at seasonality, size, hydrological connectivity, and water storage capacity of the water bodies. They are relevant indicators for the study of eco-hydrological regions with the presence of water bodies near urban spaces. From this perspective, we see the relevance of establishing a monitoring system based on Sentinel satellite data. The satellite data reveal the spatial patterns of water bodies and may be used for the evaluation and monitoring of hydric resources at the spatial and temporal scales that are adequate for urban spaces. The relationship of the water bodies with an urban area and a rural one is studied through an analysis of proximity. The population living in these areas prone to flooding was located. Due to frequent inundation, this kind of tropical zones needs to be prepared to face possible flooding and public health problems related to infectious diseases that are mainly transmitted by mosquitoes that reproduce during the rainy season.

2. Case study: state of Tabasco, Mexico

2.1. Study area

The area of study is located in the low Grijalva-Usumacinta eco-hydrological region, which was defined by the Freshwater Ecoregions of the World Map (FEOW) [16]. The FEOW defines a freshwater ecoregion as *"A large area encompassing one or more freshwater systems that contains a distinct assemblage of natural freshwater communities and species*". The Grijalva-Usumacinta ecoregion located in Tabasco México, as reported by FEOW, covers a surface of 112,008 Km^2. **Figure 1** shows the delimitation of the study area, based on FEOW. The freshwater ecoregion is coherent with the country's hydrological zones, according to works carried out by the Conagua [17]. The water body systems studied in this chapter are included in the three most important hydrological systems of the country. Conagua [17] reports an average annual precipitation of 1709 mm for the Grijalva-Usumacinta hydrological region, during the years 1971–2000. The region has a mean natural total superficial runoff of 117,396 (hm^3/year), defined by its 87 hydrological basins.

The Grijalva River flows through the city of Villahermosa, and through the state of Chiapas and small parts of the state of Campeche. The Grijalva River flows into the Usumacinta River; both rivers form the swampy delta called Pantanos de Centla, which flows into the Gulf of Mexico. The Laguna de las Ilusiones and Laguna del Negro are inner lagoons located in the city of Villahermosa.

The city of Villahermosa, as well as the hydrological systems to which it belongs, forms a large plain of fluvial origin. Normally, it is a water circulation zone. During the dry season (with precipitation <100 mm) in the period from February to April, according to SMN [18], the drastic seasonal descent of water bodies decreases food and vegetation that are vital for many species. During the intense rainy season (July to October), hurricanes and cold fronts occurring in the study area, frequently lead to flooding and landslides on highways causing

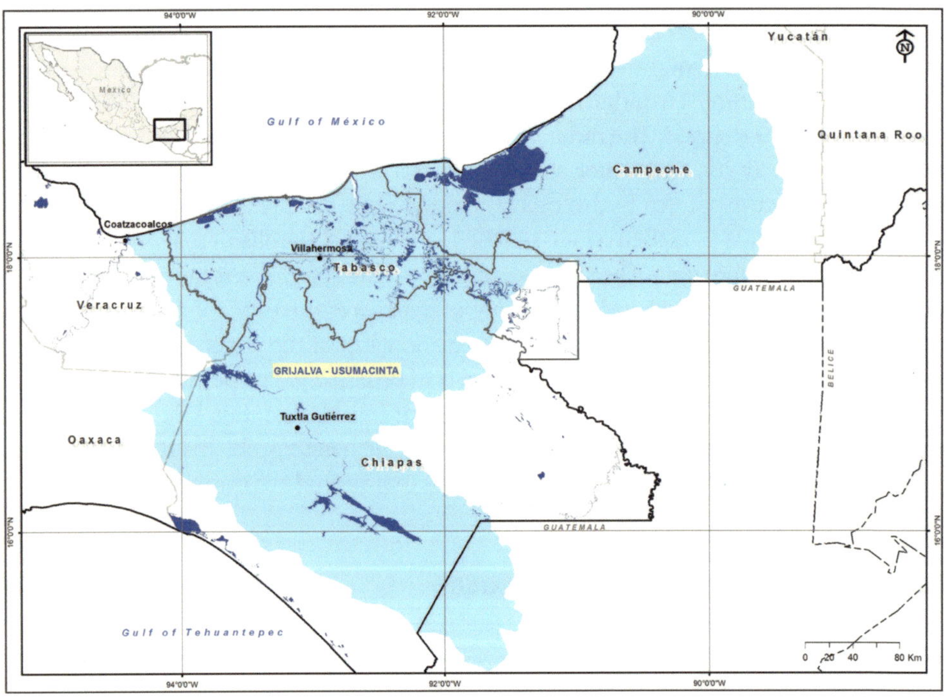

Figure 1. Map of hydrological ecoregions of the FEOW, Grijalva-Usumacinta region, in blue. The map also shows the limits of the state of Tabasco.

disasters, and leaving the populations centers isolated and without communication [19]. This is documented in historical records of flooding and atypical rainy seasons. For the state of Tabasco, the Dartmouth Flood Observatory (DFO) reports events with damages to the population during the months of October to December (October 10–17, 2000; October 12–28, 1999, and October 28, 2007–December 1, 2007). Gama et al. [19] point out that during the 2006–2010 period, the Grijalva-Usumacinta watershed had six extreme events with flooding, generating risk for the population. They also include a risk map for the whole state.

Two economic regions are identified in the state of Tabasco: the Grijalva and the Usumacinta regions. The Grijalva region is subdivided into the Chontalpa, Centro, and Sierra economic subregions, and the Usumacinta region is subdivided into the Ríos and Pantanos subregions. The Grijalva region is the most populated one in the state, with the greatest industrial and commercial growth. Here we find the most important urban centers, such as the city of Villahermosa. Loss of natural ecosystems as a result of human activities can be observed. The rural and urban localities of the Grijalva region in Tabasco have undergone rapid population growth. Urban expansion has taken place without planning processes. As a result, the population's quality of life has been affected by drought during the dry season and by high precipitation during the rainy season. Also, the inhabitants now suffer from water shortage in a region that is normally considered to have high water availability.

Figure 2 shows a map including the urban limits of the city of Villahermosa and its metropolitan area, located in the Centro municipality. In this map, the urban area of the years 2000, 2005, 2010, and 2015, as defined by the National Institute of Statistics and Geography (INEGI, Spanish acronym, Mexico) land cover data, are compared. In this 15 year period, the urban area increased by 14 km^2.

2.2. Methodology

2.2.1. SAR signal and water surfaces

Recent research has shown the feasibility of Synthetic Aperture Radar (SAR) images, as a source of information on water surface. The advantage of SAR images is that they provide useful data regardless of the weather conditions (can penetrate clouds, rain, and smoke) and lighting conditions. The radar signal is a relationship between the energy transmitted by the sensor's microwaves to the Earth and the reflected energy, that is, the return signal to the sensor. The energy or signal that returns to the sensor is called backscattering [20]. The acquisition parameters and the characteristics of the water surfaces affect the radar return signal. According to Lewis [21], the parameters affecting radar backscatter are the following: roughness, dielectric constant, incidence angle, and wavelength.

Many water bodies are related to wetlands and aquatic vegetation living submerged or partly submerged, or with leaves that float on the surface. Lewis [21] says the following about the radar response: *"Corner reflection or Dihedral reflection often is observed from the combination of trees beside a specularly reflecting surface."* In this case, the signal is enhanced and there is a double bounce effect between the trunks and water, which send most of the transmitted signal to the SAR antenna and which increases the water contrast, resulting in bright tones on SAR images [22].

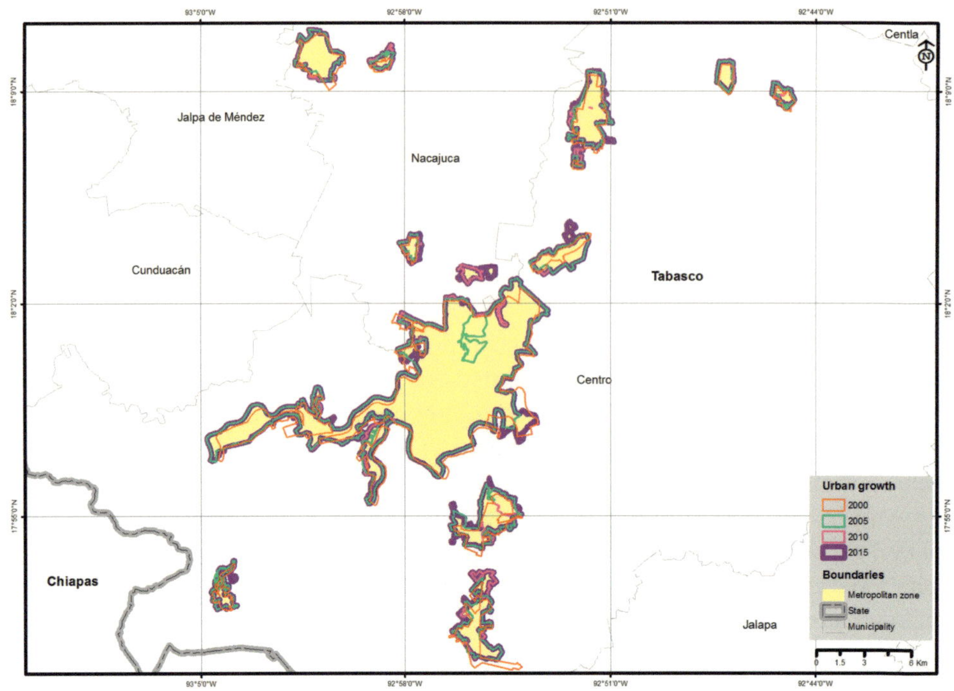

Figure 2. Regional context. Urban growth map showing urban limits and comparing the following years: 2000, 2005, 2010, and 2015, considering the area of the Villahermosa metropolitan zone in the Grijalva basin. Source: INEGI cartography and land cover data.

The backscattering coefficient term, σ^0, is a physical quantity. This coefficient is highly dynamic and is usually expressed in decibels (dB). Some refer to it as sigma naught, or sigma zero. Backscattering coefficients provide information on the observed surface, with values ranging from +5 dB for very bright objects to −40 dB for very dark surfaces. The surface of a water body is detectable in radar images. The backscattering coefficient of water, using Sentinel 1A data with VV polarizations, varies from −6 to −15 dB, and for VH polarizations, it varies from −15 to −24 dB [23].

When a lake is calm, the behavior with respect to the radar signal is called specular reflector, which means that the incident radar signal reaching the lake reflects almost all the incident energy far from the sensor. Plain surfaces like water bodies appear in dark tones in the radar image, since the antenna does not receive a strong signal. The contrast of the dark tone makes an ideal separation possible, between the water cover and the land.

When the surfaces of a lake or river have emerging vegetation and movement due to wind, rain, and vegetation, the result is that superficial layers of the water body show superficial roughness and thus, it is difficult to have contrast between the tones of the land-water covers. In terms of values of the backscattering coefficient, this means that the roughness of the water's surface results in a high signal return and the separation of the land-water covers is made more difficult because of the decrease in the contrast.

Besides, we must consider that water and land separation may be problematic if there are types of covers that may present the specular reflection effect, such as smooth surfaces (airport runways, streets, or plain bare ground), radar shadows, very dry ground (sand), and open sand dunes which appear in dark tones resulting in low backscatter returns under −20 dB [21].

The relationship between backscatter and local incidence angle is also very important to distinguish between surface conditions [23, 24]. The geometric effects are due to the terrain's relief; when the incidence angles are large, they tend to have radar shadowing effects, and on the contrary, small incidence angles favor layover effects. To decrease the effects of shadowed areas, images have been recorded at different vision angles or directions of the orbit's trajectory (ascending or descending), as reported by Chunxi et al. [25]. O'Grady et al. [26] tried to mitigate the problems related to wind effects, comparing the SAR response and using the difference between the incidence angle of the radar signal and backscattering over land and water; also, the way in which radar backscattering varies with the incidence angle may be used to differentiate the land-water separation.

Among the works using X- and C-band wavelengths to produce water surface maps, we highlight the ones referred by Bolanos et al. [27] and Brisco et al. [28, 29], with Radarsat-1 and 2 (C-band) data, Martinis and Twele [30] and Gstaiger et al. [31] with Terrasar-X (X-band) data, and Twele et al. [32], Cazals et al. [33], Muro et al. [34], and Yésou et al. [35] using Sentinel 1.

HH polarization is preferred for cartography of flooded vegetation and water classification, since it maximizes canopy penetration and increases the contrast between water and its surroundings [28, 29]. Besides, HH has less backscatter on rough water surface (roughness induced by wind), than VV, and is thus better for windy conditions.

The Sentinel-1 scene offers the potential to obtain data to describe the temporal and seasonal variations in water bodies' extension. Sentinel-1A C-band usually acquires data on VV/VH polarization over the Earth's land cover. Twele et al. [32] analyzed the polarization behavior with Sentinel data; the VV band shows a higher signal response in double bounce conditions than VH, but VH polarization shows higher backscatter variability on land surfaces. In terms of thematic accuracy, VV polarization is slightly higher than that of VH polarization.

2.2.2. *Available Sentinel data 1*

Sentinel-1A [36] carries a synthetic aperture radar (SAR) and produces, since the end of 2014, images that cover the same scene in a 12 day cycle, with a mean resolution of 20 m. The data acquisition frequency of Sentinel-1 satellites for the same geographic scene offers the potential to obtain data on the temporal and seasonal variations in the behavior of water bodies. The SAR data were obtained from the server of the Copernicus Open Access Hub. The total number of images was 36 (track 172, 99, and 26), and they were obtained between January and December 2015.

Reference Data. To determine the training sites and the validation of estimates of the water body extraction G3WBM and INEGI water body data (2015) were used. Global 3-second Water Body Map (G3WBM) (November 2015) [37] is a global scale high-resolution water body map obtained using multi-temporal Landsat data. This cartography shows the water body distribution at a global level, with a spatial resolution of 90 m.

2.2.3. Workflow of water body extraction

The analysis of data from the Sentinel 1A temporal series is based on a processing chain shown in **Figure 3**. First, corrections were carried out in a pre-processing phase. Based on these images, water body extraction was performed with a supervised classification method, using the Support Vector Machine algorithm to obtain the binary mask of the lakes, lagoons, and rivers. The validation of the obtained results was done comparing the areas of certain selected lakes with the reference data to obtain confusion matrices and overall accuracy and Kappa indexes. Finally, estimates of water frequency were based on the average monthly mean of water presence/absence. The software that was used was: (1) SNAP for pre-processing, (2) Monteverdi, for water mask extraction, and (3) gvSIG, Quantum GIS; besides, for cartographic representation of results, Quantum GIS was used for post-classification.

2.2.3.1. Data pre-processing

The Orbit State Vector in the metadata file of Sentinel C band product is a file that includes the general acquisition parameters. To ensure that the file of the precise orbit information is updated, the precise orbit ephemerides (POD) file is provided and located by the Sentinel-1 payload data ground segment (PDGS). Thermal noise removal (TNR) is used to remove noise before performing calibration. This step is needed to eliminate those values that are seriously affected by this type of noise.

The purpose of the Radiometric Calibration is to convert the digital level values to backscattering coefficient values. The SAR calibration methods proposed by Freeman [20] allow us

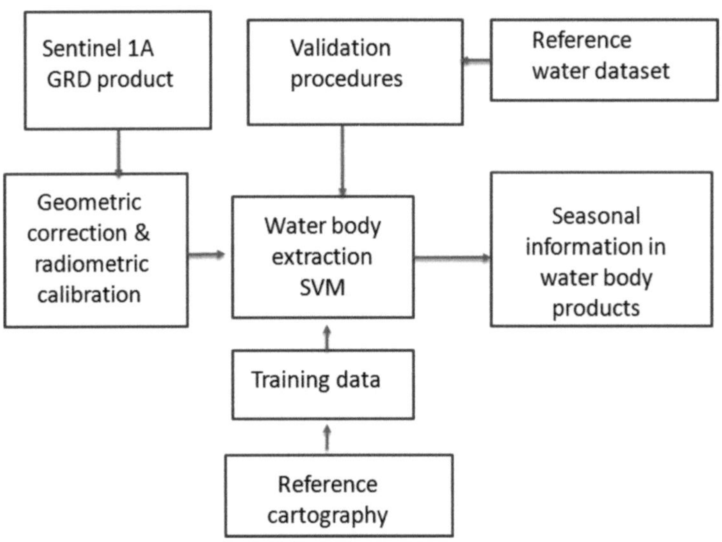

Figure 3. Study methodological scheme.

to convert radar reflectivity to physical units; these depend on the characteristic SAR system, ERS or Sentinel-1. The calibration equation of the Sentinel products follows the generic ENVISA-ASAR expression. According to the Sentinel-1 (S-1) Instrument Processing Facility (IPF) document [38], the equation of the σ^0 radar cross-section is simplified to: $\sigma^0 = DN^2/(A^2_{dn}k)*\sin(\alpha)$, where DN is the pixel Digital Number, k is the calibration constant, and A_{dn} is the product final scaling from GRD (A_{dn} is an area normalization factor and α is the incident angle). For GRD products a constant offset is also applied by the following simplified equation: $\sigma^0 = DN^2/A^2_{\sigma}$, where σ^0 is the backscattering coefficient and A_{σ} is used to transform the radar reflectivity into radar cross-section σ^0. The σ^0 values can be transformed to dB values considering the following formula: $\sigma^0_{db} = 10\log_{10}\sigma^0$.

The multi-look refers to the divisions of the total aperture, which is subdivided into several sub-apertures. With each sub-aperture, an image is formed. All images are averaged. The typical number of looks is 1–4. Each sub-aperture gives an independent look to the illuminated scene. Each one of these looks will be subjected to speckle, but by the sum and average of resolution cells, all together they will form the resulting radar image with a reduced speckle. The multi-look process is done during data acquisition. The average can be obtained in the frequency domain by averaging images of individual frequency parts in range and/or azimuth; or in the spatial (time) domain by averaging individual resolution cells in range and/or azimuth [39]. The multi-look process allows for the creation of squared pixels.

In general, all radar images show a degree of speckle that is manifested as a salt and pepper granular pattern. Actually, speckle is a form of noise that degrades the quality of an image and can make visual interpretation more difficult. Noise in SAR images has a structure that is different from other image processing systems. The ideal filter used to reduce speckle is the one that does not lose any information. For example, in the homogenous areas, the filter must preserve the radiometric information, as well as preserve the edges of the different areas in the textured zones. In practice, Lee Sigma Filter was used to filter speckle. Finally, Geometric Terrain Correction was performed. The main geometric distortions are: (1) layover, (2) shadowing, and (3) foreshortening. The algorithm uses a Digital Elevation Model SRTM 3-arc-second to make the correction. The resulting corrected image remains in its correct geographic orientation. The final coordinates are geographical WGS84.

2.2.3.2. Surface water extraction

To date, automated algorithms have been developed for pattern recognition, detection, delineation, and segmentation of water bodies. This progress entails pre-processing of images, obtaining backscattering coefficients, and tackling problems related to body delineation caused by shadows, wind, aquatic vegetation, humidity gradients, and sediments, as previously mentioned.

There are many studies that have been done using radar satellite images to detect surface water. The method that is most broadly reported in the literature is the intensity thresholding method applied to the SAR image. In this method, water is separated from land in the intensity images, but the accuracy of the results is based on the ability to differentiate the land pixels vs. water in the intensity domain. There are several modifications to the thresholding method: [29, 32, 35, 40–44]. There are also different ways to improve the thresholding method. As an example, Cazals et al. [33] use a hysteresis thresholding algorithm which is complemented with data from in situ piezometric measurements combined with a Digital Terrain Model derived from LiDAR data.

With respect to the detection of changes and segmentation algorithms to determine the pixels belonging to water, we refer to Gstaiger [31], Kuenzer [45], Matgen [46], and Muro [34]. There are other methods such as split-based automatic thresholding and classification refinement and the hybrid context-based model, combining causal with non-causal Markov image modeling on irregular hierarchical graphs, applied by Martinis [30, 47], respectively. Twele [32] uses a fuzzy-logic-based classification refinement, with Sentinel-1 ground range detected (GRD) data. Lopez-Caloca et al. [48] used Sentinel-1 (GRD) data to classify the Support Vector Machine method (SVM).

We chose the SVM method to classify the set of monthly images. Post-classification was carried out by means of the filter with QGIS tools, to eliminate unclassified points, decrease the granular effect, and eliminate loose pixels of both classes. To make the database more coherent, each water layer was examined by visual interpretation.

2.2.3.3. Validation

To obtain an estimate of the error related to classifications. The results were compared based on the Kappa index. The comparison study of the area was performed for 2015 with the G3WBM and INEGI database information sources. The results of the water body binary masks obtained from the ascending mode were re-scaled to a spatial resolution of the 90 m reference product. To carry out the validation process, the classes from the compared images were homogenized in the following categories: permanent water bodies, flood areas, and other classes. Our results were considered as permanent water for pixels with frequency higher than 73%.

2.2.3.4. Frequency indicator

The performed dynamic study is based on the work of Andreoli [43], and the determination of the frequency indicator. The frequency time $P_y(w)$ is an average of the monthly mean of water surfaces $P_{y,m}(w)$ expressed in percent of year (Eq. (1))

$$P_{y,m}(w) = \left(\frac{1}{N_{y,m}} \sum_{t=1}^{N_{y,m}} w_{y,m,t} \right) x100 \tag{1}$$

where m is the month of the year, y is the year, $N_{y,m}$ is the number of water extent layers of the month (m) of the year (y) number of pixels identified as water ($N_{y,m} = 12$), $w_{y,m,t}$ is the pixel value of the water extent layer extracted from Sentinel data for the month (m) of the year (y); with $t \in [1;n_m]$ and $w_{(m,t)} \in \{0,1\}$, and t is the resolution of the data.

2.2.4. Results

The inundation frequency map is shown in **Figure 4**. **Tables 1** and **2** present the results from the validation of the classifications based on confusion matrices and the kappa index. As can be seen,

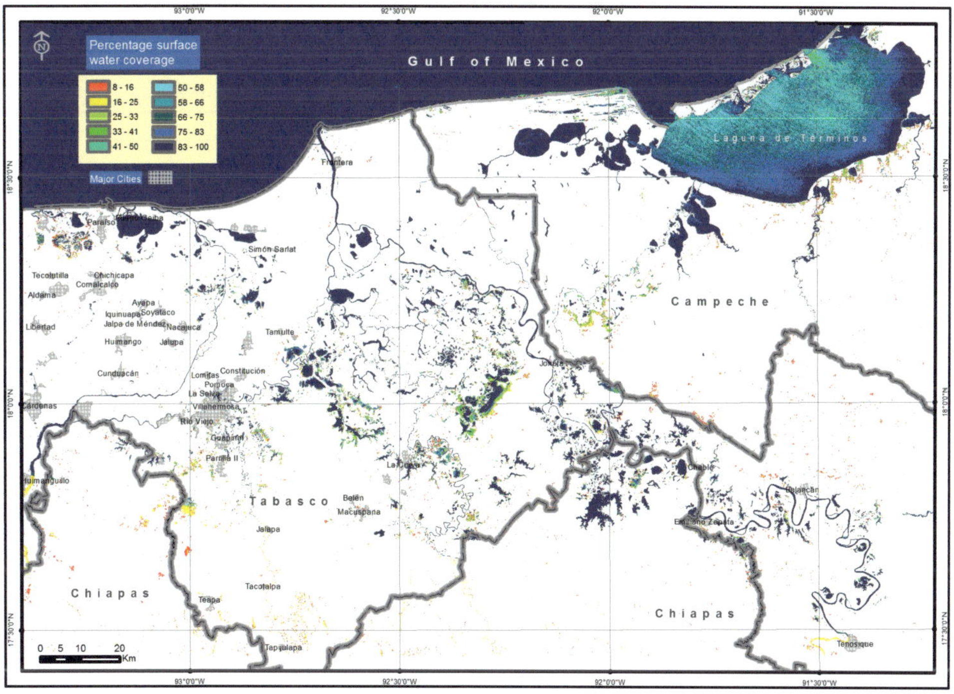

Figure 4. Inundation frequency map for 2015, the map including cities of interest in the state of Tabasco.

		Reference data				
		Non-water	Water	Totals	Commission error (%)	Estimated kappa
Predicted class	Non-water	172,937	12,875	185,812	6.9	0.88
	Water	2416	229,618	232,034	1.0	0.97
	Totals	175,353	242,493	417,846		
	Omission error (%)	1.4	5.3			

Table 1. Confusion matrix of the water/non-water classification using Sentinel radar data (March 2015) and adjusted INEGI's 2015 land cover data as reference data.

the kappa was in all cases very high, 0.92 for the comparison against INEGI data, and 0.97 for the comparison against G3WBM water body maps. The results show that it is feasible to use the water body extraction method with the SVM algorithm to monitor the whole state of Tabasco.

In global numbers for Tabasco state, the water body surface obtained from Yamazaki, based on the G3WBM product, is 1,531,429.40 km^2, which is slightly lower than the number obtained from this study's map, which is 1,620,868.01 km^2. **Figure 4** shows a map with the inundation frequency by pixel, obtained by monthly binary masks of 2015. The dark blue color shows permanent water bodies and the red areas show the increase in water due to flooding.

		Reference data				
		Non-water	Water	Totals	Commission error (%)	Estimated kappa
Predicted class	Non-water	292,013,286	3,230,447	295,243,733	1.1	0.97
	Water	5,943,121	301,431,080	307,374,201	1.9	0.96
	Totals	297,956,407	304,661,527	602,617,934		
	Omission error (%)	1.9	1.06			

Overall accuracy: 98.47%, Kappa Statistic (standard): 0.97.

Table 2. Confusion matrix of the water/non-water classification using Sentinel radar data (March 2015) and G3WBM version 1.2 water body maps as reference data.

Figure 5 shows a close-up of the extension of water bodies during 2015, over the Villahermosa metropolitan area. In the maps, permanent water cover is represented by blue and the non-permanent water cover by red. To visualize the proximity of urban and rural centers, the administrative delimitation of these areas with human settlements is included. There, we may see the natural expansive process of the water bodies during the study year.

To complete the analysis, we carried out analysis of areas of influence based on distance to urban centers, calculating buffers using the Euclidian distance algorithm. **Figure 6** shows the

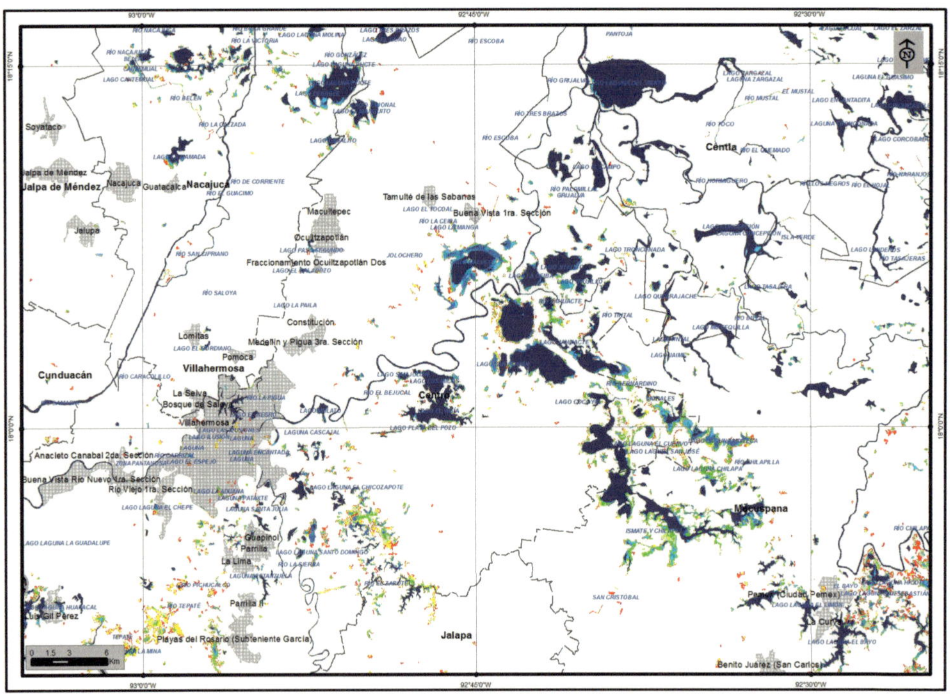

Figure 5. Villahermosa metropolitan area in the Centro municipality.

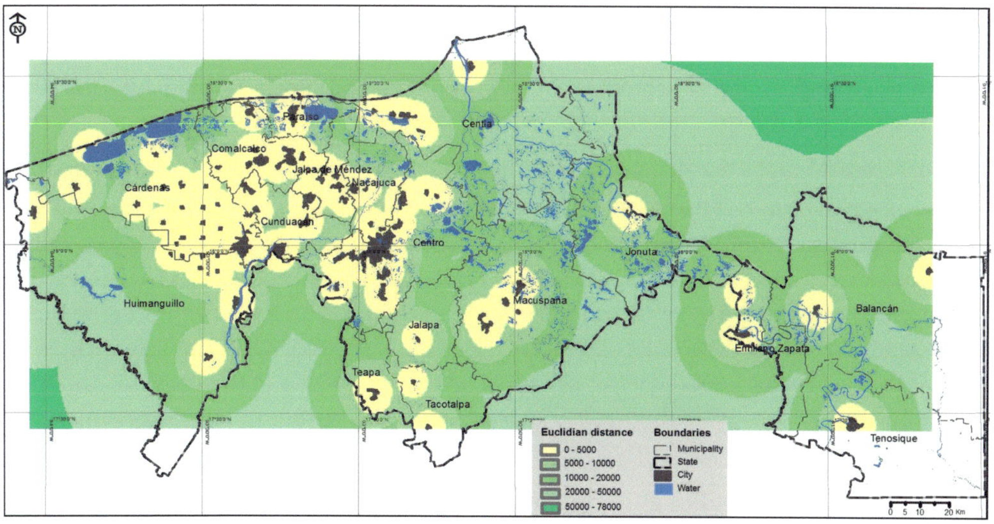

Figure 6. Map of urban and rural centers and areas of influence (buffers) of water bodies to 5000 m, detected by the Sentinel-1A in 2015. The map shows the binary mask of the water bodies.

location of the urban and rural centers on the map, with their closeness to the water bodies. Given that urban and rural centers influence land-use and water bodies the presence of, we see that Villahermosa, Macuspana, Emiliano Zapata, Comacalco, and other municipalities are located near several water bodies, which can make them susceptible to flood related risks. This type of analysis allows us to make information available that can help local communities manage water related risk. They get knowledge about the behavior of their nearest neighboring water bodies in terms of increases or decreases in their extension and storage capacity.

Sentinel missions are reducing the observation limitations of other satellite systems and strengthening the ability to monitor changes in the planet. In this case, we had a study zone, in which cloud cover is a normal condition. Therefore, this kind of zones can only be studied using radar satellite systems. The qualitative change in production and management of data from these missions is an opportunity to have spatial-temporal information that fills knowledge gaps and which will allow us develop more trustworthy models and predictions of the behavior of the Earth's ecosystems. It will also help us generate timely information for the management of flood risks, earthquakes, and other natural phenomena. The free availability of data from Sentinel satellites is the main condition to advance in a rapid way in research and in the prediction of changes in the Earth's resources and their implications for the economy and the safety of societies.

3. Conclusion

To monitor resources, we need to develop studies on the specific ecosystems in each region. The new technologies, such as the Sentinel missions, give us valuable information for superficial water body monitoring. The supervised classification technique was adequate, as shown by the results section. The procedure had very good performance (kappa >0.9, and overall accuracy >90%). This is relevant considering that it was applied to a flatland area with a dynamic hydrological regime composed of many lakes and rivers formed by rainfall and runoff. Using this technique, the seasonal changes could be detected.

This study contributes to the characterization of superficial water bodies, using short term time series, and help determine whether there are changes in their extension. It also means the beginning of the process to get to know some geographical relationships, such as the proximity of human settlements and freshwater bodies than can pose risks to the safety and economy of the communities living there. The results of this study show the potential for monitoring superficial water bodies, providing basic information for planning land and water management as well as other activities such as flood control programs. Also, an analysis of the proximity of human settlements to superficial water bodies is important for the diagnosis of socio-economic aspects that can affect safety, economy, and health, which are related to increasing poverty levels.

Superficial water body studies at the regional and local levels are increasingly important to understand processes as climate change and its effects on the Earth's surface. With short- and long-term satellite time series studies of water bodies, it will be possible to determine if there are changes related to natural variability or other changes related to man-made damages. This will allow us to develop trustworthy models and predictions of the behavior of the Earth's ecosystems that can help us develop better territorial planning processes.

Acknowledgements

This work was realized within the framework of Geo-Intelligence Laboratory (CentroGeo). Authors would also like to thank Dr. Dai Yamazaki for G3WBM information.

Author details

Alejandra A. López-Caloca[1*], Felipe Omar Tapia-Silva[2] and Guadalupe Rivera[1]

*Address all correspondence to: alopez@centrogeo.edu.mx

1 Centro de Investigación en Geografía y Geomática "Ing. Jorge L. Tamayo" A.C. (CentroGeo), México

2 Universidad Autónoma Metropolitana Iztapalapa (UAM-I), Lab. Geomática Aplicada a Recursos Naturales, Dep. de Hidrobiología, México

References

[1] Moggridge HL, Hill MJ, Wood PJ. Urban aquatic ecosystems: The good, the bad and the ugly. Fundamental and Applied Limnology. 2014;**185**(1):1-6. DOI: 10.1127/fal/2014/0716

[2] Lambin EF, Geist HJ, Lepers E. Dynamics of land-use and land-cover change in tropical regions. Annual Review of Environment and Resources. 2003;**28**:205-241. DOI: 10.1146/annurev.energy.28.050302.105459

[3] Alberti M, Booth d Hill K, Avolio C, Coe S, Spirandelli D. The impact of urban patterns on aquatic ecosystems: An empirical analysis in Puget lowland sub-basins. Landscape and Urban Planning. 2007;**80**:345-361. DOI: 10.1016/j.landurbplan.2006.08.001

[4] Musaoglu N, Tanik A. Identification of land-cover changes through image processing and associated impacts on water reservoir conditions. Environmental Management. 2005;**35**(2):220-230. DOI: 10.1007/s00267-003-0270-4

[5] Steele MK, Heffernan JB. Morphological characteristics of urban water bodies: Mechanisms of change and implications for ecosystem function. Ecological Applications. 2014;**24**:1070-1084. DOI: 10.1890/13-0983.1

[6] Bernhardt ES, Palmer MA. Restoring streams in an urbanizing world. Freshwater Biology. 2007;**52**:738-751. DOI: 10.1111/j.1365-2427.2006.01718.x

[7] Kucukmehmetoglu M, Geymen A. Measuring the spatial impacts of urbanization on the surface water resource basins in Istanbul via remote sensing. Environmental Monitoring and Assessment. 2008;**142**:153-169. DOI: 10.1007/s10661-007-9917-6

[8] Dunne T. Urban hydrology in the tropics: Problems, solutions, data collection, and analysis. Urban climatology and its applications with special regard to tropical areas. TR Oke, World Meteorol. Org, WMO. Mexico D.F. 1986;**652**:405-434

[9] Yue J. Urban rivers: A landscape ecological perspective. Hidrology Current Research. 2012;**3**:125. DOI: 10.4172/2157-7587.1000125

[10] Steele MK, Heffernan JB, Bettez N, et al. Convergent surface water distributions in U.S. cities. Ecosystems. 2014;**17**:685. DOI: 10.1007/s10021-014-9751-y

[11] Rodríguez RE. Las lagunas continentales de Tabasco. Universidad Juárez Autónoma de Tabasco. Villahermosa: Tabasco Mexico; 2012. p. 145-166 ISBN 968-7991-31-3

[12] Deng Y, Jiang W, Tang Z, Li J, Lv J, Chen Z, Jia K. Spatio-temporal change of Lake water extent in Wuhan urban agglomeration based on Landsat images from 1987 and 2015. Remote Sensing. 2017;**9**:270. DOI: 10.3390/rs9030270

[13] GWP Global Water Partnership. Integrated Water Resources Management. Technical Advisory Committee (TAC) Background; 2000. Paper no. 4 Stockholm, Sweden. 24 p

[14] Savenije HHG, Van der Zaag P. Integrated water resources management. Concepts and issues. Physics and Chemistry of the Earth. 2008;**33**:290-297. DOI: 10.1016/j.pce.2008.02.003

[15] Guzinski R, Kass S, Huber S, Bauer-Gottwein P, Jensen IH, Naeimi V, Doubkova M, Walli A, Tottrup C. Enabling the use of earth observation data for integrated water resource management in Africa with the water observation and information system. Remote Sensing. 2014;**6**:7819-7839. DOI: 10.3390/rs6087819

[16] Abell R, Thieme M, Revenga C, Bryer M, Kottelat M, Bogutskaya N, Coad B Mandrak N, Contreras-Balderas S, Bussing W, Stiassny MLJ, Skelton P, Allen GR, Unmack P, Naseka A, Sindorf RNg N, Robertson J, Armijo E, Higgins J, Heibel TJ, Wikramanayake E, Olson D, Lopez HL, REd R, Lundberg JG, Sabaj Perez MH, Petry P. Freshwater ecoregions of the world: A new map of biogeographic units for freshwater biodiversity conservation. Bioscience. 2008;**58**:403-414. DOI: 10.1641/B580507

[17] Atlas del Agua en México 2015. Secretaría de Medio Ambiente y Recursos Naturales y Comisión Nacional del Agua. 2015; 20-22, 56, 105 p. Printed and made in Mexico. www.conagua.gob.mx

[18] Servicio Meteorológico Nacional (SMN). Available from: http://smn.cna.gob.mx/es/informacion-climatologica-ver-estado?estado=tab

[19] Gama L, Ortiz-Pérez MA, Moguel-Ordoñez E, Collado-Torres R, Díaz-López H, Villanueva-García C, Macías-Valadez ME. Flood risk assessment in Tabasco, Mexico. Water Resources Management. 2011;**VI**:631-639. DOI: 10.2495/WRM110561

[20] Freeman A. SAR calibration: An overview. IEEE Transactions on Geoscience and Remote Sensing. 1992;**30**(6). DOI: 10.1109/36.193786

[21] Lewis AJ. Geomorphic and hydrologic applications of active microwave remote sensing. In: Henderson FM, Lexis AJ, editors. Principles and Applications of Imaging Radar. N.Y, The United States of America: Wiley and Sons; 1998. p. 567-629 ISBN-13: 978-0471294061

[22] Brian B. Remote sensing of wetlands: Applications and advances. In: Tiner RW, Lang MW, Klemas VV, editors. In Chapter 7 in Remote Sensing of Wetlands: Applications and Advances. NY, USA: CRC Press; 2015. p. 119-136 ISBN: 978-1-4822-3735-1

[23] Manjusree P, Kumar LP, Bhatt CM, Rao GS, Bhanumurthy V. Optimization of threshold ranges for rapid flood inundation mapping by evaluating backscatter profiles of high incidence angle SAR images. Int. J. Disaster Risk Sci. 2012;**3**(2):113-122. DOI: 10.1007/s13753-012-0011-5

[24] O'Grady D, Leblanc M, Gillieson D. Relationship of local incidence angle with satellite radar backscatter for different surface conditions. International Journal of Applied Earth Observation and Geoinformation. 2013;**24**:42-53. DOI: 10.1016/j.jag.2013.02.005

[25] Chunxi X, Jixian Z, Guoman H, Zheng Z, Jiao W. Water body information extraction from high resolution airborne synthetic aperture radar image with technique of imaging in different directions and object-oriented. Proceeding of the ISPRS Congress Silk Road for Information from Imagery, Beijing. 2008:165-168 http://www.isprs.org/proceedings/XXXVII/congress/7_pdf/2_WG-VII-2/18.pdf

[26] O'Grady D, Leblanc M, Bass A. The use of radar satellite data from multiple incidence angles improves surface water mapping. Remote Sensing of Environment. 2014;**140**:652-664. DOI: 10.1016/j.rse.2013.10.006

[27] Bolanos S, Stiff D, Brisco B, Pietroniro A. Operational surface water detection and monitoring using radarsat 2. Remote Sensing. 2016;**8**:285. DOI: 10.3390/rs8040285

[28] Brisco B, Touzi R, Van Der Sanden JJ, Charbonneau F, Pultz TJ, D'Iorio M. Water resource applications with RADARSAT-2 a- preview. International Journal of Digital Earth. 2008;**1**(1):130-147. DOI: 10.1080/17538940701782577

[29] Brisco B, Short N, Van der Sanden JJ, Landry R, Raymond D. A semi-automated tool for surface water mapping with RADARSAT-1. Canadian Journal of Remote Sensing. 2009;**35**(4):336-344. DOI: doi.org/10.5589/m09-025

[30] Martinis S, Twele A. A hierarchical spatio-temporal markov model for improved flood mapping using multi-temporal X-band SAR data. Remote Sensing. 2010;**2**(9):2240-2258. DOI: 10.3390/rs2092240

[31] Gstaiger V, Huth J, Gebhardt S, Wehrmann T, Kuenzer C. Multi-sensoral and automated derivation of inundated areas using TerraSAR-X and ENVISAT ASAR data. International Journal of Remote Sensing. 2012;**33**(22):7291-7304. DOI: 10.1080/01431161.2012.700421

[32] Twele A, Cao W, Plank S, Martinis S. Sentinel-1-based flood mapping: A fully automated processing chain. International Journal of Remote Sensing. 2016;**37**(13):2990-3004. DOI: 10.1080/01431161.2016.1192304

[33] Cazals C, Rapinel S, Frison P-L, Bonis A, Mercier G, Mallet C, Corgne S, Rudant J-P. Mapping and characterization of hydrological dynamics in a coastal marsh using high temporal resolution sentinel-1A images. Remote Sensing. 2016;**8**(7):570. DOI: 10.3390/rs8070570

[34] Muro J, Morton C, Florian R, Strauch A, Thonfeld MF, Gunter, Conradsen K, Nielsen AA, Hüttich C, Skriver H. Short-term change detection in wetlands using sentinel-1 time series. Remote Sensing. 2016;**8**:10-795. DOI: 10.3390/rs8100795

[35] Yésou H, Huber C, Haouet S, Lai X, Huang S, Fraipont P, Desnos Y-L. Exploiting sentinel 1 time series to monitor the largest fresh water bodies in PR China, the Poyang Lake. IEEE International Geoscience and Remote Sensing Symposium, IGARSS. 2016:3882-3885. DOI: 10.1109/IGARSS.2016.7730008

[36] European Space Agency (ESA), Sentinel-1 User Handbook. 2013. https://sentinel.esa.int/documents/247904/685163/Sentinel-1_User_Handbook

[37] Yamazaki D, Trigg MA, Ikeshima D. Development of a global ~90 m water body map using multi-temporal Landsat images. Remote Sensing of Environment. 2015;**171**:337-351. DOI: 10.1016/j.rse.2015.10.014

[38] European Space Agency (ESA). Radiometric Calibration of S-1 Level-1 Products Generated by the S-1 IPF (Sentinel-1 Instrument Processing Facility IPF) https://sentinel.esa.int/documents/247904/685163/S1-Radiometric-Calibration-V1.0.pdf

[39] Raney RK. Radar fundamentals: Technical perspective. In: Henderson FM, Lexis AJ, editors. Principles and Applications of Imaging Radar. N.Y: Wiley and Sons; 1998. p. 31-32 ISBN-13: 978-0471294061

[40] White L, Brisco B, Pregitzer M, Tedford B, Boychuk L. RADARSAT-2 beam mode selection for surface water and flooded vegetation mapping. Canadian Journal of Remote Sensing. 2014;**40**:135-151. DOI: 10.1080/07038992.2014.943393

[41] Li J, Wang S. An automatic method for mapping inland surface waterbodies with Radarsat-2 imagery. International Journal of Remote Sensing. 2015;**36**(5):1367-1384. DOI: 10.1080/01431161.2015.1009653

[42] Martinis S, Twele A, Voigt S. Towards operational near real-time flood detection using a split-based automatic thresholding procedure on high resolution TerraSAR-X data. Natural Hazards and Earth System Sciences. 2009;**9**:303-314. DOI: 10.5194/nhess-9-303-2009

[43] Andreoli R, Yésou Y, Li J, Desnos Y-L, Shifeng H, De Fraipont P. Poyang Hu (Jiangxi Province, P. R. of China) area variations between January 2004 and June 2006 using ENVISAT low and medium resolution time series. Geographic Information Sciences. 2007;**13**(1-2). DOI: 10.1080/10824000709480629

[44] Sabel D, Naeimi V, Greifeneder F, Wagner W. Investigating radar time series for hydrological characterisation in the lower Mekong Basin. In: Chapter 17, Remote Sensing Time Series, Volume 22 of the Series Remote Sensing and Digital Image Processing. Switzerland: Springer International Publishing. 2015;**22**:357-381. DOI: 10.1007/978-3-319-15967-6. ISBN: 978-3-319-15966-9

[45] Kuenzer C, Guo H, Huth J, Leinenkugel P, Li X, Dech S. Flood mapping and flood dynamics of the Mekong delta: ENVISAT-ASAR-WSM based time series analyses. Remote Sensing. 2013;**5**(2):687-715. DOI: 10.3390/rs5020687

[46] Matgen P, Hostache R, Schumann G, Pfister L, Hoffmann L, Savenije HHG. Towards an automated SAR-based flood monitoring system: Lessons learned from two case studies. Physics and Chemistry of the Earth. 2011;**36**(7-8):241-252. DOI: 10.1016/j.pce.2010.12.009

[47] Martinis S, Clandillon S, Twele A, Huber C, Plank S, Maxant J, Cao W, Caspard M, May S. Improving the Extraction of Crisis Information in the Context of Flood, Landslide, and Fire Rapid Mapping Using SAR and Optical Remote Sensing Data. Vol. 18. Vienna, Austria, 17-22-04-2016: EGU 2016; 2016 EGU2016-4928

[48] López-Caloca AA, Tapia-Silva FO, Rivera G. In: Ouwehand L, editor. Living Planet Symposium, Proceedings of the Conference Held 9-13-05-2016 in Prague, Czech Republic. ESA-SP. 2016;**740**. p. 144. ISBN: 978-92-9221-305-3

Chapter 9

Identifying Changes in Trends of Summer Air Temperatures of the USA High Plains

Denis Mutiibwa, Ayse Kilic and Suat Irmak

Additional information is available at the end of the chapter

http://dx.doi.org/10.5772/intechopen.71788

Abstract

Change in climate variables, especially air temperature, can substantially impact water availability, use, management, allocation, and projections for rural and urban applications. This study presents analyses for detecting summer air temperature change by investigating trends of two separate climate-periods in the USA High Plains. Two trend periods, the *reference period* (1895–1930) and the *warming period* (1971–2006), were investigated using parametric and nonparametric methods. During the reference period, minimum air temperature (T_{min}) was statistically stationary at a nonsignificant increasing rate of 0.02°C/year. However, from early 1970s, T_{min} increased at a significant rate of 0.02°C/year. The maximum air temperature (T_{max}) had a weaker warming signal than T_{min} during the reference period. During the warming period, T_{max} had a cooling trend at a nonsignificant rate of −0.004°C/year. About 22% of the High Plains had significant warming trends before 1930. Compared to the summers before 1930, the summer temperatures of the High Palins since the 1970s increased, on average, by 0.86°C. Overall, parametric methods lead to the conclusion that 50% of the study area experienced a significant warming trend in T_{min}. In comparison, nonparametric methods indicated that 94% of the study area experienced a warming trend. Overall, in recent decades, summer average temperatures in the High Plains have been warming as compared to the early twentieth-century decades, and the warming is most likely driven primarily by increasing nighttime T_{min}.

Keywords: climate change, water resources, air temperature, USA High Plains, parametric and nonparametric tests, Kendall tau, generalized linear models

1. Introduction

The United Kingdom Meteorological Office Hadley Centre/Climate Research Unit (HadCRUT3) [1], the United States National Climatic Data Center (NCDC) [2], and the

Goddard Institute for Space Studies (GISS) [3] regularly monitor and update global and hemispheric surface temperature changes. The three institutions use similar data; however, they employ different interpolation techniques and 30-year climatological (base) periods (HadCRUT: 1961–1990, NCDC: 1901–2000, and GISS: 1951–1980) to construct global surface temperature anomalies. Despite the differences in methods, all three groups produced similar global temperature anomalies for the period of 1880–2009 [4]. These anomalies show evidence of a warming trend in temperatures over the period of 1880–2009. The global temperature anomalies presented by NCDC [4] and GISS [3] reveal three trend periods of distinct means and trends in global temperatures: the period of 1880–1930 with the mean value of −0.21°C; the period of 1935–1975 with the mean of about 0.0°C; and the period of 1976–2006 that has a warming trend of about 0.02°C/year. The HadCRUT global temperature anomalies have a similar profile; however, due to a variant base period of 1961–1990, the reference level (zero line) is shifted higher and the three mean values are offset from GISS and NCDC.

While aforementioned analyses provide global averages, changes in trends and magnitudes of climatic variables can vary between the regions. Thus, it is important to identify these changes locally, which can provide important and useful information on a variety of topics, including agricultural science and practices, hydrologic analyses, climate change studies, etc. The rationale of this study was to investigate the potential changes in regional temperatures starting from the early 1970s, by referencing that period to an earlier period with the least observed changes (warming) since the advent of temperature monitoring and archiving in the late nineteenth century. The two periods investigated are (i) 1895–1930 as the *reference period* and (ii) 1971–2006 as the *warming period*. The assumption of the reference period is that, as indicated in the NCDC, GISS, and HadCRUT data, this period has the least warming signal in measured temperature records, which started in the late nineteenth century. The years between the reference period and the warming period (1936–1965) were used as the 30-year climatological (base) period. This base period was selected as the independent period for estimating the anomalies of the reference period and warming period. The study was conducted on summer temperatures that were computed as the average of June, July, and August monthly temperatures.

Other studies [5–9] estimated trends in air temperatures and other variables, and their uncertainties, using various statistical methods to derive and test the trends and presented extrapolation methods used. Each statistical method can have advantages or disadvantages, depending on application characteristics and objectives of the study. Folland et al. [7] presented one of the earliest global and hemispheric surface warming trends that attempted to quantify the major sources of uncertainty. They calculated the global and hemispheric annual temperature anomalies by combining land surface air temperature and sea surface temperatures. They observed that the best linear fit to annual global surface temperature showed an increase of 0.61 ± 0.16°C between 1861 and 2000.

Time-dependent variables such as temperature have a probable confounding effect of serial autocorrelation that may violate statistical assumptions in some of the trend studies that fit simple ordinary least squares. This is because the standard errors of serially correlated variables are typically underestimated. Santer et al. [10] corrected the effects of serial correlation by adjusting the sample size to an equivalent sample size, which is regarded as the number of effectively independent observations in a sample. However, the student's *t*-test adjusted by the equivalent sample size performed poorly, because the equivalent sample size was poorly estimated and it was incorrectly assumed that the adjusted statistic has a student's *t* distribution under the null hypothesis [11].

In this study, parametric and nonparametric statistical methods are used to investigate trends in the air temperature series. The parametric method used the autoregressive process in the time series to fit generalized least squares that are superior to simple ordinary least squares. Autoregressive processes are used in climate research since they provide approximations of discretized ordinary linear differential equations subject to stochastic forcing [12]. For the nonparametric method, the Kendall tau trend analysis was applied in this study. The advantage of nonparametric methods is that the statistical assumption of normality in the data is not strictly necessary. The two statistical methods were combined to assess trends in the minimum air temperature (T_{min}), maximum air temperature (T_{max}), and mean air temperature (T_{mean}). This study has three specific objectives: (i) to characterize potential trend differences between the reference period and warming period in the summer temperatures of the High Plains, (ii) to identify the temperature variable with the warmest trend, and (iii) to analyze the performance of parametric and nonparametric statistical methods in characterizing trends in temperature series in terms of mean anomalies.

2. Materials and methods

2.1. Input data

Monthly data for T_{max}, T_{min}, and T_{mean} for 204 weather stations across the High Plains were obtained from the United States Historical Climatology Network (USHCN) (**Figure 1**). Data from all 204 stations were complete for the entire study period (1896–2006). Thus, the spatial sampling error was assumed homogenous for the entire study period. The USHCN data were downloaded from the NCDC [6] (http://www.ncdc.noaa.gov), which maintains, distributes, and conducts data quality checks. Inhomogeneity and missing data are typically caused by changes in instrumentation, measuring techniques, station location, observation frequency, and environment shifts due to relocations [13, 14]. The NCDC ensures good-quality data by subjecting the data to a comprehensive quality control, inhomogeneity correction, and removal of all monthly mean outliers that differed from their climatology by more than 2.5 standard deviations [15–17]. The NCDC homogenizes the dataset to remove impacts of urban warming and other artifacts on measured temperature [15, 18]. In the case of missing data, a network of surrounding stations is used to interpolate the missing values, thus producing a continuous data series.

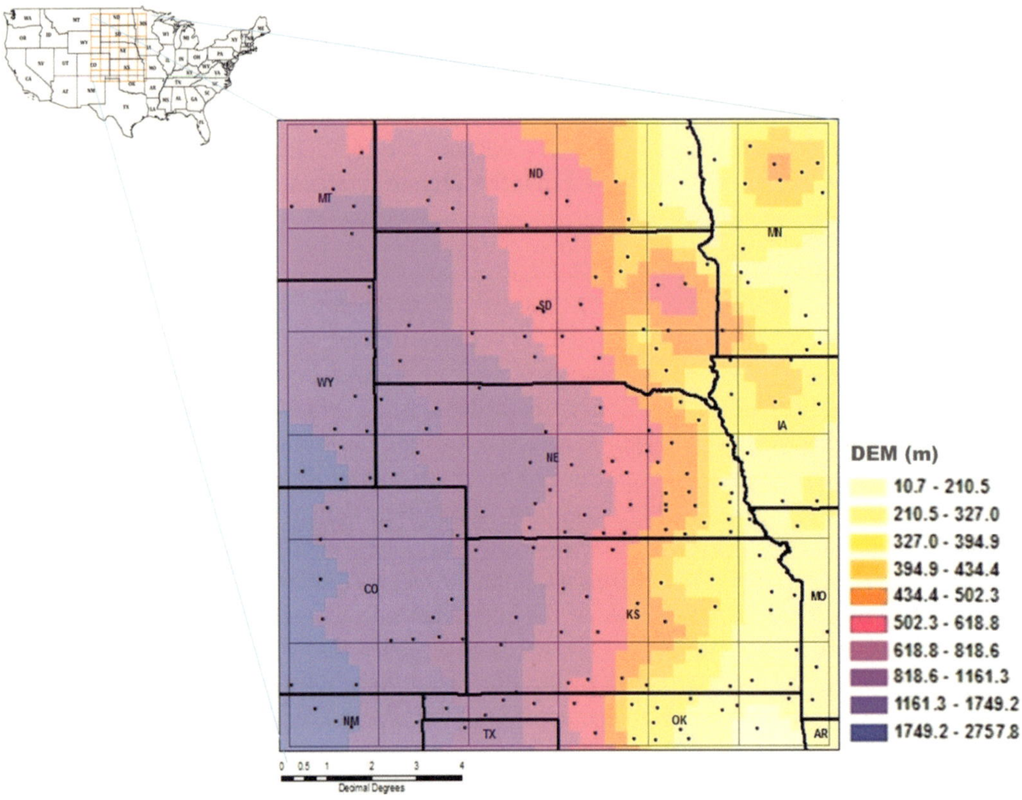

Figure 1. A map of the study region, the US High Plains, showing the locations of the weather stations (black dots) used in the study, and the elevation across the region in meters. DEM: Digital Elevation Model.

2.2. Spatial domain and gridding

The High Plains region of the central United States extends over 12 states: Nebraska, Kansas, and South Dakota as the main states of interest, surrounded by buffer areas of North Dakota, Minnesota, Wyoming, Montana, Colorado, Iowa, Missouri, Oklahoma, and Texas. The regional climate is described as middle-latitude dry continental climate with abundant sunshine, moderate precipitation, frequent winds, low humidity, and high evaporation rates [19]. This study divides the region into 36 grids of 2 degrees in size. On average, there were six weather stations in every grid. Studies have shown that monthly temperature anomalies are mainly a function of large-scale circulation patterns [20]. Therefore, the number of stations required to describe the monthly anomalies over an area is comparatively less extensive [20]. For instance, for a global terrestrial coverage at a 5-degree grid, estimates indicated that the effective number of independent stations at a monthly timescale was about 100 well-spaced sampling sites [21].

2.2.1. Creating anomalies and gridding

The monthly anomalies of T_{mean}, T_{min}, and T_{max} were computed based on the climate anomaly method (CAM) [22]. The period from 1936 to 1965 was considered the base period for determining the norm monthly temperatures for June, July, and August. The monthly anomalies at each station were derived as the difference between the actual monthly temperature and the respective mean monthly norm temperature. The summer anomalies were then computed as the average of the three monthly anomalies (June, July, and August) for each year. By averaging the summer anomalies of stations in each grid, a 2-degree grid of summer temperature anomalies was created over the entire High Plains region. For long-term average spatial variability across the region (**Figure 2**), the summer temperatures in each grid were averaged for the entire study period (1895–2006).

2.3. Parametric trend analysis

The generalized linear model (GLM) was used to estimate linear trends in the temperature series. The serial autocorrelation in the residual term was accounted for using the autoregressive error model, as described below. Considering a time series of summer temperature anomalies, y_t, from one of the grids in **Figure 1**, GLM estimates the trend (β) in the series as follows:

$$y_t = \alpha + t\beta + \omega_t, \quad t = l, \cdots n_t \tag{1}$$

The term ω_t represents the residuals that contain a deterministic process and a random process (errors) as:

$$\varepsilon_t = \omega_t\left(1 - \varphi_1 B - \varphi_2 B^2 - \cdots \varphi_p B^p\right) \; ; \quad \varepsilon_t \sim IN(0, \sigma^2) \tag{2}$$

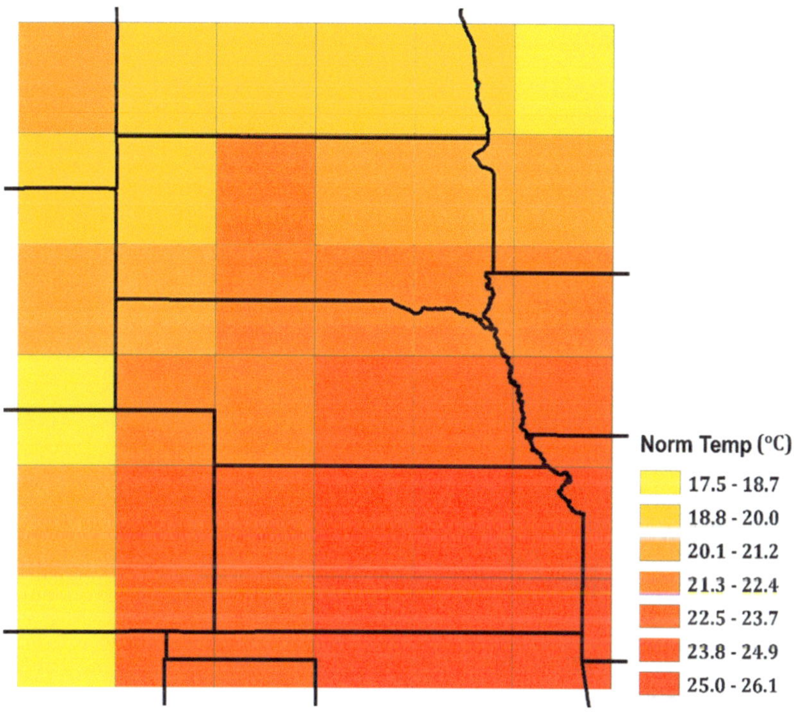

Figure 2. Spatial patterns of normal temperature across the US High Plains for the period from 1895 to 2006 in the summer months (June, July, and August).

where α is the intercept; n_t is the number of study years in a trend period (35 years); ϵ_t is the error term, which is assumed independent (I) and normally (N) distributed with a mean of 0 and a variance of $\varphi_p\sigma^2$; φ_p is the autoregressive error model parameter; p is the autoregressive order; and B is the backshift operator. The right-hand side of Eq. (2) represents the deterministic process imbedded in the residuals of Eq. (1). By removing the deterministic process from the residual term, the autoregressive error model generates independent and normally distributed errors that are critical for the maximum likelihood estimation of trend (β). The β estimates from the GLM are thus the unbiased and minimum variance estimators of trend in the temperature series. A parametric t-test was used to test the null hypothesis that the trend in the temperature series was zero at 0.05 level of significance. The t-statistic was computed as a quotient of estimated trend (β) and its standard error (S_β).

2.4. Nonparametric trend analysis

The Kendall tau [23], a nonparametric method, was used to determine trends in absolute values of average summer temperatures. The advantage of Kendall tau method over GLM is that the statistical assumption of normality is not strictly necessary and the test statistics are less impacted by outliers in the temperature series. This method has been used to compute trends in climatic and hydrological series [24, 25]. The method is actually applied on ranks of the absolute values; therefore, Kendall tau estimates are relative measures of strength and direction of actual trends. The trend estimates (Kendall tau) were tested at the 0.05 significance level.

3. Results and discussion

3.1. Spatial variation of summer temperatures in the High Plains

The summer norm temperatures of the High Plains in **Figure 2** are averages of 111 years (1895–2006). There is an expected north to south increasing trend in summer temperatures across the region. This spatial variability is driven by the cool air masses from the arctic in the northern part of the Plains and the warm air masses from the Gulf of Mexico in the southern part of the Plains. In the south of the region, there is another west to east increasing trend, which is associated with the elevation gradient across the Plains. Elevation across the Plains increases from less than 200 m in the east to more than 2700 m in the west (**Figure 1**). In the southwest of the Plains, another phenomenon caused by the Chinook winds is associated with the cooler summer temperatures in that region. These high westerly winds subject the Rocky Mountain range to periodic severe turbulence in summer. The winds have a strong cold frontal passage downslope; however, the winds eventually undergo an adiabatic warming process as they move eastward [26]. The effect of Chinook winds can be felt over 100 miles from the Rock Mountains before dissipating as the winds mix and saturate with the atmosphere.

3.2. Temporal variation of summer temperatures in the High Plains

The High Plains region T_{mean} anomalies (base period, 1935–1965) from 1895 to 2006, along with a 6-year moving average of the anomalies, and the Northern Hemisphere Land (NHL) summer temperature anomalies (base period, 1901–2000) are presented in **Figure 3**. There is higher variability in High Plains summer anomalies compared to the NHL because of increase in internal climate variability with decrease in regional size [27]. High Plains anomalies exhibit most of the region's extreme events, including droughts (e.g., 1930s Dust Bowl era), extreme warm and cool summers of the study period. The NHL anomalies are smoother, because the extreme events in one part of the region are likely smoothed by moderate conditions in other paths of the region.

Prior to 1930, **Figure 3** shows that the summer temperature anomalies were mostly below zero. The coolest summer of the study period was in 1915 at 3.44°C below the base period (1935–1965) norm. In the mid-1930s, the Plains experienced the worst drought of the study period. The drought was characterized by high temperatures, high winds, and low rainfall [28]. Temperatures in the drought years of 1934 and 1936 still hold the record of the top two hottest summers in the High Plains. The NHL anomalies do not explicitly feature the drought, an indication that the entire northern hemisphere was not in the drought during the 1930s. The drought peaked in 1936 and declined in the 1940s; however, summer temperatures never reached to pre-1930 conditions. Even though we did not calculate the trend and its significance, from the early 1940s to the 1960s, there appears to be an overall cooling, although there are years with positive anomalies (i.e., several consecutive years in the 1950s). According to Ref. [29], this cooling was observed across entire North America, lasting from 1945 to 1976.

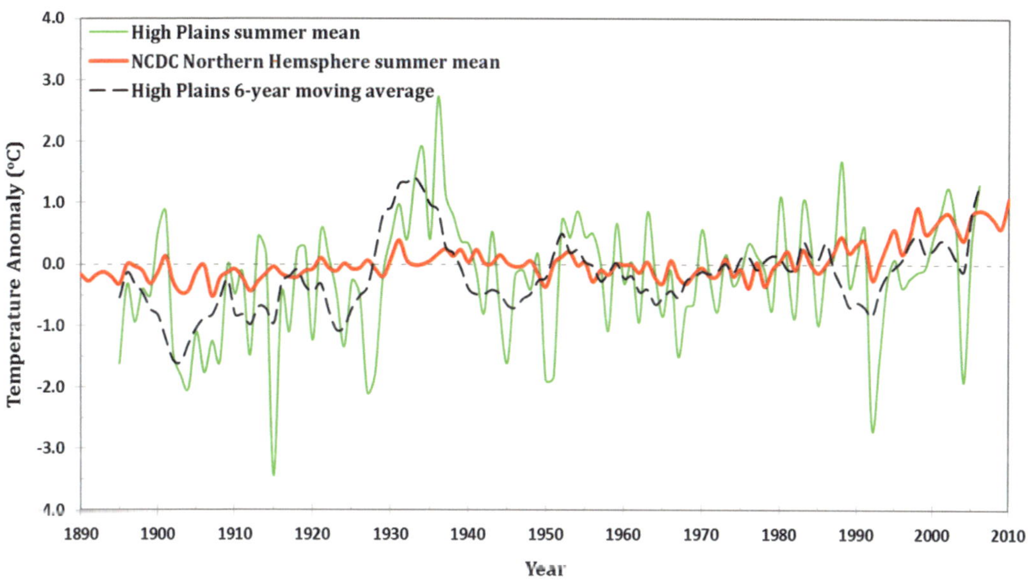

Figure 3. Comparison of variability and trends in the summer mean air temperature (T_{mean}) anomalies relative to 1935–1965 base period for the High Plains (solid green line) and the Northern Hemispheric summer land temperature anomalies relative to 1901–2000 base period (source: NCDC/NESDIS/NOAA, solid red line), and the 6-year moving average (dashed black line) summer T_{mean} anomalies relative to 1935–1965 base period for the High Plains region of the USA.

Since 1970s, the NHL anomalies indicate a warming trend that could be attributed to global warming, primarily due to increases in greenhouse gas emissions. In the High Plains anomalies, the warming trend is obscured by high variability in the series; however, the moving average elucidates the warming trend. Compared to the summers before 1930, the summer temperatures of the High Palins since the 1970s increased, on average, by 0.86°C. In a global study, Ref. [4] similarly observed that temperatures in the first decade of the twenty-first century were about 0.80°C warmer than the beginning of the twentieth century (1880–1920).

3.3. Trends in T_{min}

3.3.1. Parametric analysis

The maximum likelihood estimates of T_{min} trends for the two trend periods are presented in **Figure 4(a** and **b)**. During the reference period (**Figure 4a**), nine grids (25% of the High Plains) had significant warming trends. This warming trend occurred mainly in the northern part of the Plains. The trends were insignificant in the south part of the region. Across the High Plains, there is no statistically significant evidence of a trend in T_{min}.

Figure 4b shows the spatial trend patterns in T_{min} during the warming period. More area (50%) across the region, still mostly in the north, experienced significant warming in T_{min} during the warming period. Unlike the reference period, the overall T_{min} trend of the region in the warming period was significant at 0.02°C/year.

3.3.2. Nonparametric analysis

The nonparametric Kendall tau estimates of trend patterns in T_{min} are shown in **Figure 5(a** and **b)** (the trend estimates in **Figures 5**, **7**, and **9** are tau values, which are relative estimates

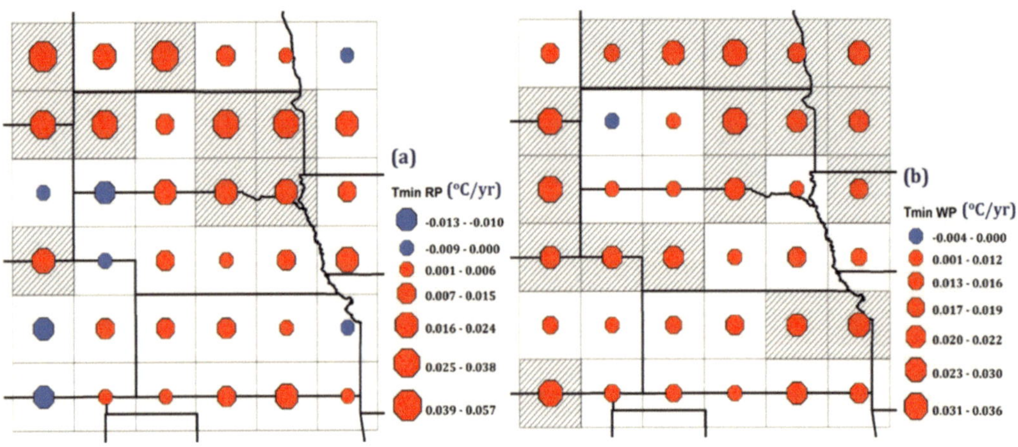

Figure 4. Spatial trend patterns of summer T_{min} for (a) *reference period* (1895–1930) and (b) *warming period* (1971–2006) computed from generalized linear models. The hatching shows areas where the trend was significant. The red and blue indicate warming and cooling trend effects, respectively. The trend estimates are in degrees Celsius per year. RP: reference period; WP: warming period.

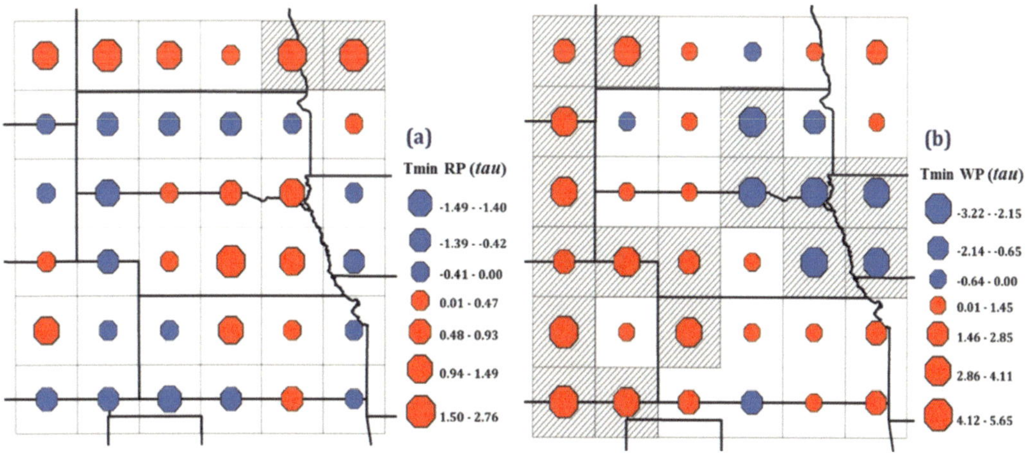

Figure 5. Spatial trend patterns of summer T_{min} for (a) *reference period* (1895–1930) and (b) *warming period* (1971–2006) computed by Kendall tau method. The hatching shows areas where the trend was significant. The red and blue indicate warming and cooling trend effects, respectively. The trend estimates are in degrees Celsius per year. RP: reference period; WP: warming period.

of strength and direction of the actual trends. These values describe the spatial variability of the trends across the region during the specified trend period). The Kendall tau analysis detected similar significant trend patterns as GLM in the northern part of the Plains during the reference period. About 22% of the High Plains had significant warming trends before 1930. The results from nonparametric and parametric analyses are comparable in identifying trend patterns during the reference period. During the warming period, however, Kendall tau test was more effective in detecting significant trends across the region (**Figure 5b**). According to Kendall tau, most of the region experienced significant warming during the warming period. The nonparametric test is effective in terms of identifying significant trends, because of its relative insensitivity to extreme values in the series. Both parametric and nonparametric analyses on T_{min} agreed that during the reference period the overall trend in T_{min} was stationary. And, since 1971, T_{min} has significantly increased at the rate of 0.02°C/year. Overall, parametric methods lead to the conclusion that 50% of the study area experienced a significant warming trend in T_{min}. In comparison, nonparametric methods indicated that 94% of the study area experienced a warming trend.

3.4. Trends in T_{max}

3.4.1. Parametric analysis

Figure 6(a and b) shows the maximum likelihood estimates of T_{max} trends for the two trend periods. The trends in T_{max} were mostly weaker than trends in the T_{min}. In **Figure 6a**, the trend patterns in reference period were only significant in four grids of the northern High Plains (11% of the High Plains). The magnitudes of the overall trend of the entire High Plains region during the reference period were insignificant at a rate of 0.01°C/year.

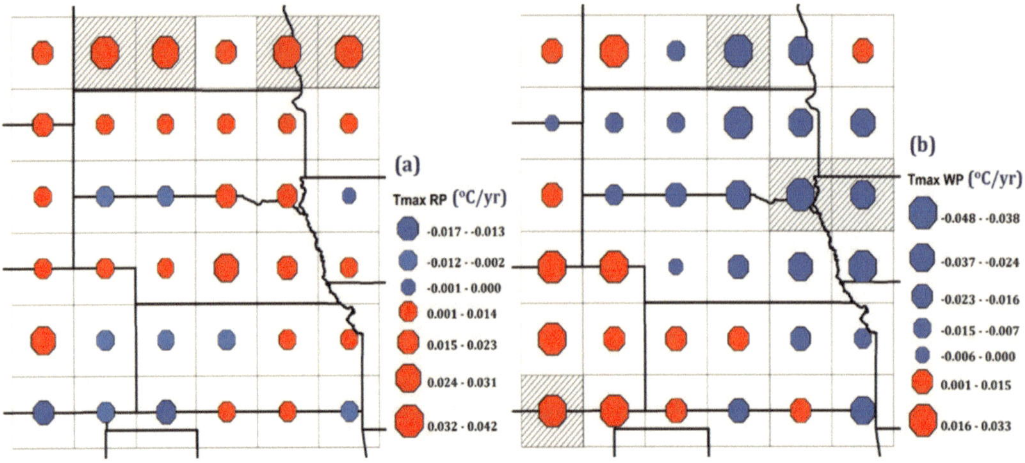

Figure 6. Spatial trend patterns of summer T_{max} for (a) *reference period* (1895–1930) and (b) *warming period* (1971–2006) computed from generalized linear models. The hatching shows areas where the trend was significant. The red and blue indicate warming and cooling trend effects, respectively. The trend estimates are in degrees Celsius per year. RP: reference period; WP: warming period.

During the warming period (**Figure 6b**), only one grid in the southwest of the High Plains had a significant warming trend in T_{max}. The other areas with significant trends had cooling effects. For the rest of the region, the trends were insignificant and many had a cooling effect. In fact, the overall trend in T_{max} during the warming period was insignificant with a cooling effect of −0.004°C/year. Folland et al. [7] also observed that in the last quarter of the twentieth century the Central United States cooled by 0.2–0.8°C in summer. While the significance is identified

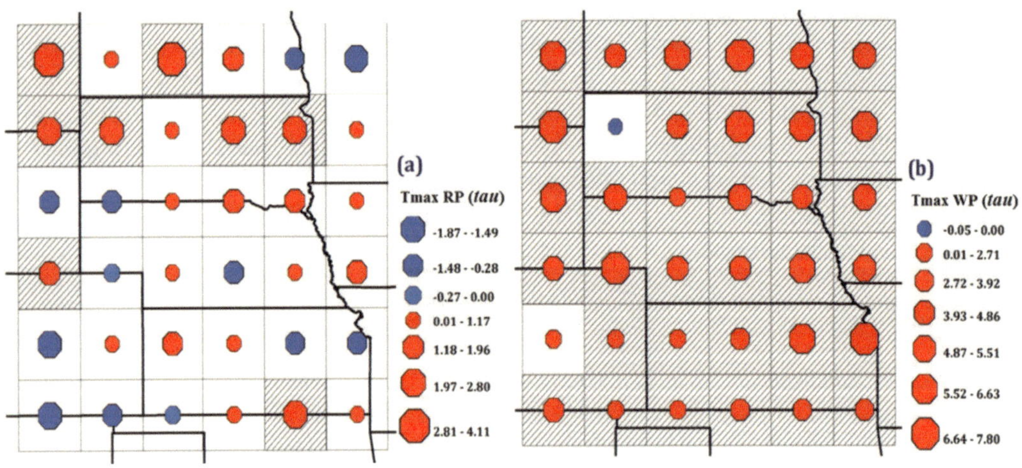

Figure 7. Spatial trend patterns of summer T_{max} for (a) *reference period* (1895–1930) and (b) *warming period* (1971–2006) computed by Kendall tau method. The hatching shows areas where the trend was significant. The red and blue indicate warming and cooling trend effects, respectively. The trend estimates are in degrees Celsius per year. RP: reference period; WP: warming period.

statistically in this and other studies, in practice, the error of measurement of Tair with various thermometers, including mercury thermometers used in historical datasets, is probably in the ±0.5°C range and, thus, a trend of −0.004°C/year might have a "statistical" significance but may not be truly a "physical significance," depending on the measurement resolution of the thermometers used.

3.4.2. Nonparametric analysis

The trend patterns in T_{max} determined by Kendall tau method are shown in **Figure 7(a** and **b)**. During the reference period, the Kendall tau trend patterns (**Figure 7a**) were similar to the GLM trend patterns (**Figure 6a**), with significant warming only observed in the northeast of the region. For the warming period (**Figure 7b**), more significant warming trends were detected by Kendall tau method, especially in the western part of the Plains. In the central-eastern part, the Plains experienced a significant cooling during the warming period (**Figure 7b**). As with the parametric method, the nonparametric overall trends in T_{max} during the reference and warming periods were also insignificant. A possible influence on T_{max} during the warming period is evaporative cooling from extensive irrigation practice in the High Plains during the summer months of June, July, and August. However, we have not conducted analyses to explicitly study whether irrigation practice is the cause of the trends. In addition to irrigation practices, other practices such as elevations; changes in land use, population, management practices, changes in the locations and surroundings, as well as instrumentation used in the weather stations from which climate were obtained, etc. can also influence trends and magnitudes in air temperatures, which were not considered in this study.

The potential impact(s) of land use (e.g., irrigation) impact on surface air temperature has been studied primarily using large-scale climate models with varying results. For example, Ref. [30] used a regional climate model, which showed the regional irrigation cooling effect (ICE) exists, opposite in sign to urban heat island effects. The magnitude of the ICE has strong seasonal variability, causing large dry-season decreases in monthly mean and maximum temperatures, but little change in rainy season temperatures. Their model produced a negligible effect on monthly minimum temperature. In California, the modeled regional ICE is of similar magnitude, but opposite sign, to predictions for future regional warming from greenhouse gases. Given their modeling results for California and the global importance of irrigated agriculture, they concluded that past expansion of irrigated land has likely affected observations of surface temperature, potentially masking the full warming signal caused by greenhouse gas increases.

Lara et al. [31] reported the seasonally varying temperature responses of four regional climate models (RCMs)—RSM, RegCM3, MM5-CLM3, and DRCM—to conversion of potential natural vegetation to modern land cover and land use over a 1-year period. Three of the RCMs supplemented soil moisture, producing large decreases in the August mean (−1.4 to −3.1°C) and maximum (−2.9 to −6.1°C) 2-m air temperatures where natural vegetation was converted to irrigated agriculture. Conversion to irrigated agriculture also resulted in large increases in relative humidity (9–36% absolute change). Modeled changes in the August minimum 2-m air temperature were not as pronounced or consistent across the models. Converting natural vegetation to urban land cover produced less pronounced temperature effects in all models, with the magnitude of the effect dependent upon the preexisting vegetation type and urban parameterizations. Their modeling results indicated that, overall, the RCM results indicate that the temperature impacts of land-use change are most pronounced during the summer months, when surface heating is strongest and differences in surface soil moisture between irrigated land and natural vegetation are largest.

Using ensemble simulations, Ref. [32] evaluated the impacts of irrigation changes on air temperatures in the twentieth century. Simulation results indicated that early in the century, irrigation was primarily localized over southern and eastern Asia, leading to significant cooling in boreal summer (June-August) over these regions. This cooling spread and intensified by the century's end, following the rapid expansion of irrigation over North America, Europe, and Asia. Irrigation also led to boreal winter (December-February) warming over parts of North America and Asia in the latter part of the century, due to enhanced downward longwave fluxes from increased near-surface humidity. They suggested, based on their modeling results, these trends reveal the varying importance of irrigation-climate interactions and suggest that future climate studies should account for irrigation, especially in regions with unsustainable irrigation resources. Lobell et al. [33] observed trends in Tmax were negative in irrigated areas of California and Nebraska, which they attributed to increase in latent heat flux and associated reduction in sensible heat flux. Irrigation development in the High Plains increased substantially over the last five decades. The irrigated area in the High Plains states increased from 8 million ha in 1980 to more than 13.4 million ha in 2000. In Nebraska, the total irrigated area has more than doubled in the last four decades, increasing from 1.6 million ha in 1970 to over 3.6 million ha in 2008 [34]. Other neighboring states also experienced considerable increases in irrigated acreage.

3.5. Trends in T_{mean}

3.5.1. Parametric analysis

In terms of trends in T_{mean} (**Figure 8a** and **b**), during the reference period, much of the High Plains did not experience significant trends, except the northern region (**Figure 8a**). For the rest of the Plains, the trend patterns are insignificant and many have a cooling effect. The overall trend during the reference period was stationary at a rate of 0.01°C/year. During the warming period (**Figure 8b**), with the exception of the western part of the plains, the rest of the region experienced insignificant trends. The overall warming period trend in T_{mean} was stationary at a rate of 0.01°C/year.

3.5.2. Nonparametric analysis

Trend patterns in T_{mean} obtained from the nonparametric procedure are shown in **Figure 9(a and b)**. During the reference period (**Figure 9a**), the results were similar to GLM findings, and the northwestern part of the Plains was the only area experiencing significant warming before the 1930s. The trend patterns during warming period that are presented in **Figure 9b** show that Kendall tau detected more significant trends in T_{mean} than GLM. Areas in east-central,

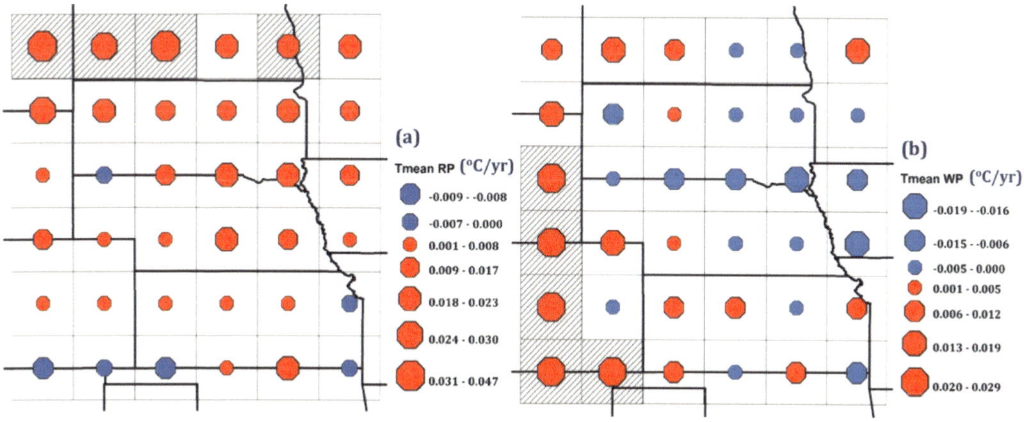

Figure 8. Spatial trend patterns of summer T_{mean} for (a) *reference period* (1895–1930) and (b) *warming period* (1971–2006) computed from generalized linear models. The hatching shows areas where the trend was significant. The red and blue indicate warming and cooling trend effects, respectively. The trend estimates are in degrees Celsius per year. RP: reference period; WP: warming period.

which had significant cooling trends in T_{max} (**Figure 7b**), had insignificant trends in T_{mean}, regardless of significant warming trends in T_{min} (**Figure 5b**). This suggests that the trends in T_{mean} are interactively influenced by the direction and magnitude of trends in T_{min} and T_{max}. For instance, some grids in the east of the region had a significant warming trend in T_{min} (**Figure 5b**) and significant cooling trend in T_{max} (**Figure 7b**), which resulted in an insignificant trend in T_{mean} (**Figure 9b**). Likewise, in the northern part of the High Plains, significant warming trends in T_{min} (**Figure 4b**), coupled with insignificant and significant cooling trends in T_{max} (**Figure 6b**), resulted in insignificant trends in T_{mean} (**Figure 8b**). Given that a significant trend

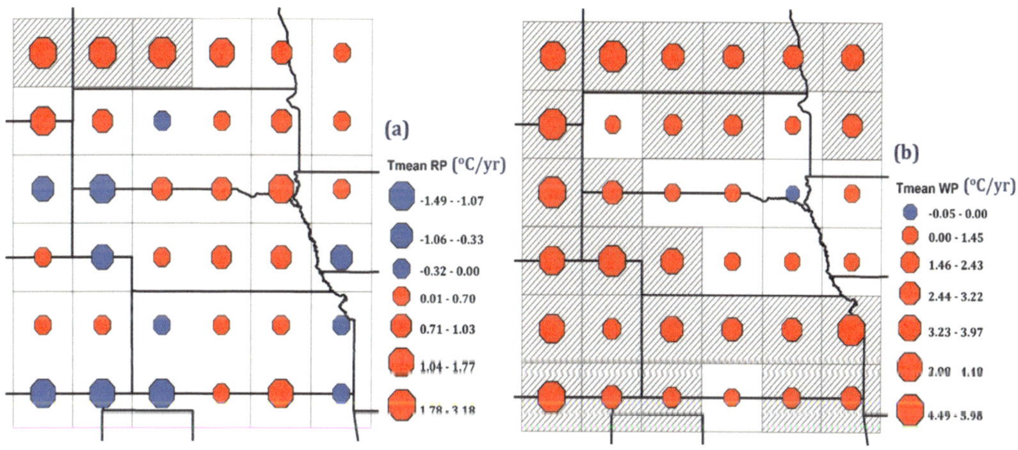

Figure 9. Spatial trend patterns of summer T_{mean} for (a) *reference period* (1895–1930) and (b) *warming period* (1971–2006) computed by Kendall tau method. The hatching shows areas where the trend was significant. The red and blue indicate warming and cooling trend effects, respectively. The trend estimates are in degrees Celsius per year. RP: reference period; WP: warming period.

in either T_{min} or T_{max} can be muted by the direction or strengthen of trend in the other, T_{mean} may be confounding to interpret and may not be an effective or ideal variable to investigate in climate change studies. In fact, Ref. [33] suggest that studies that assess impacts of climate change using only projections of T_{mean} risk over- or underestimation of uncertainties when considering process that respond differently to day and nighttime temperature.

4. Summary and conclusions

Change in climate variables, especially air temperature, can have significant impact(s) on water availability, use, management, allocation, and projections for rural and urban applications as temperature is one of the primary drivers of evaporative losses in urban and rural areas. Thus, understanding the climate change impact(s) on air temperature can aid in water use projections and other water availability assessments in urban and rural areas on large scales. We investigated trends in air temperatures of the US High Plains region in two trend periods: reference period (1895–1930) and warming period (1971–2006). Separating the data records into reference and nonreference, we think, is a unique aspect of this study that investigated long-term trends. The trend patterns were examined in T_{min}, T_{max}, and T_{mean} using parametric and nonparametric methods. The parametric method was beneficial in determining the absolute measure and direction of the trends and the nonparametric method was more effective in testing the significance of the trends. Studies that use two or more methods (parametric and nonparametric) to investigate climate parameters, especially on large scales, such as US High Plains, are useful as they may provide insight in terms of identifying the strengths and/or weakness of each method and, eventually, to rank the appropriate one. In this study, the parametric method was beneficial in determining the absolute measure and direction of the trends, but the nonparametric method was more statistically powerful in testing the significance of the trends.

The warming trends in T_{min} were stronger than in T_{max} and T_{mean}. From both trend methods, T_{min} had the biggest contrast between the reference period and the warming period. The overall trend in T_{max} over the High Plains during the warming period had an insignificant cooling effect. The trends from T_{mean} were confounded to be able to interpret since they interactively depended on the directions of trends in T_{min} and T_{max}. Both parametric and nonparametric methods showed that T_{min} was stationary during the reference period and significant warming during the warming period. In the warming period, the overall trend in T_{min} was significantly increasing by 0.02°C/year. One of the uncertainties or shortcomings of this study could be that the potential reasons of warming or cooling trends in air temperatures (e.g., changes in land use and land management, urbanization, irrigation development, and other factors) were not investigated. Further research is needed to investigate the potential interrelationships between temperature trends and change(s) in land surface characteristics on large scales. Another shortcoming of the study could be that the study focuses only on summer air temperature trends and magnitudes. However, the changes in spring, fall, and winter temperatures can also impact urban and rural water balances, especially in terms of dormant season evaporative losses, and investigating spring, fall, and winter temperature trends can be beneficial in several aspects of change in temperature impacts on urban and agricultural practices.

http://dx.doi.org/10.5772/intechopen.71788

Acknowledgements

This study is based on the work that is supported by the National Institute of Food and Agriculture, U.S. Department of Agriculture, Hatch Project, under the Project Number NEB-21-155. The study was also partially funded by the Nebraska Environmental Trust (NET) under the project number 13-146. The authors express their appreciation to USDA-NIFA and NET.

Author details

Denis Mutiibwa[1], Ayse Kilic[2] and Suat Irmak[3]*

*Address all correspondence to: sirmak2@unl.edu

1 United States Department of Agriculture, Agricultural Research Service, Beltsville, MD, USA

2 School of Natural Resources and Civil Engineering Department, University of Nebraska-Lincoln, Lincoln, NE, USA

3 Department of Biological Systems Engineering, University of Nebraska-Lincoln, Lincoln, NE, USA

References

[1] Brohan P, Kennedy JJ, Harris I, Tett SFB, Jones PD. Uncertainty estimates in regional and global observed temperature changes: A new data set from 1850. Journal of Geophysical Research. 2006;**111**:D12106. DOI: 10.1029/2005JD006548

[2] Smith TM, Reynolds RW, Peterson TC, Lawrimore J. Improvements to NOAA's historical merged land-ocean surface temperature analysis (1880-2006). Journal of Climate. 2008;**21**:2283-2293

[3] Hansen J, Sato M, Ruedy R, Lo K, Lea DW, Medina-Elizade M. Global temperature change. Proceedings of the National Academy of Sciences. 2006;**103**:14288-14293

[4] Hansen J, Ruedy R, Sato M, Lo K. Global surface temperature change. Reviews of Geophysics. 2010;**48**. DOI: RG4004

[5] Livezey R, Chen W. Statistical field significance and its determination by Monte Carlo techniques. Monthly Weather Review. 1983;**111**:46-59

[6] Wilks D. Resampling hypothesis tests for autocorrelated fields. Journal of Climate. 1997;**10**:65-83

[7] Folland CK, Rayner NA, Brown SJ, Smith TM, Shen SSP, Parker ED, Macadam I, Jones PD, Jones RN, Nicholls N, Sexton DMH. Global temperature change and its uncertainties since 1861. Geophysical Research Letters. 2001;**28**(13):2621-2624

[8] Ventura V, Paciorek CJ, Risbey JS. Controlling the proportion of falsely rejected hypotheses when conducting multiple tests with climatological data. Journal of Climate. 2004; **17**:4343-4356

[9] Livezey RE, Vinnikov KY, Timofeyeva MM, Tinker R, van den Dool HM. Estimation and extrapolation of climate normals and climatic trends. Journal of Applied Meteorology and Climatology 2007;**46**:1759-1776

[10] Santer BD, Wigley TML, Boyle JS, Gaffen D, Hnilo DJJ, Nychka D, Parker DE, Taylor KE. Statistical significance of trends and trend differences in layer-average atmospheric temperature time series. Journal of Geophysical Research. 2000;**105**:7337-7356

[11] Zwiers FW, Storch VH. Taking serial correlation into account in tests of the mean. Journal of Climate. 1995;**8**:336-351

[12] Storch VH, Zwiers FW. Statistical Analysis in Climate Research. New York: Cambridge University Press; 1999

[13] Bradley RS, Kelly PM, Jones PD, Diaz FH, Goodess C. A climatic data bank for the Northern Hemisphere, 1851-1980. In: DoE Technical Report No TR017 USDE. Washington, DC: Carbon Dioxide Research Division; 1985

[14] Parker DE. Effects of changing exposure of thermometers at land stations. International Journal of Climatology. 1994;**14**:1-31

[15] Peterson TC, Vose RC. An overview of the Global Historical Climatology Network temperature data base. Bulletin of the American Meteorological Society. 1997;**78**(12):2837-2849

[16] Peterson HD, Jones PD. Initial selection of a GCOS surface network. Bulletin of the American Meteorological Society. 1997;**78**:2145-2152

[17] Peterson TC, Vose RS, Schmoyer R, Razuvaëv V. Quality control of monthly climate data: The GHCN experience. International Journal of Climatology. 1997;**16**:1197-1226

[18] Karl TR, Williams NC, Quinlan FT. United States Historical Climatology Network (HCN) serial temperature and precipitation data, ORNL/CDIAC-30, NDP-019/R1. Oak Ridge, Tenn: Carbon Dioxide Inf. Anal. Cent., Oak Ridge Natl. Lab., U.S. Dep. Of Energy; 1990

[19] Gutentag ED, Heimes FJ, Krothe NC, Luckey RR, Weeks JB. Geohydrology of the High Plains aquifer in parts of Colorado, Kansas, Nebraska, New Mexico, Oklahoma, South Dakota, Texas, and Wyoming. USGS Professional Paper: 1400-B. 1984. 63 pp

[20] New MG, Hulme M, Jones PD. Representing Twentieth-Century Space–Time Climate Variability. Part I: Development of a 1961-90 Mean Monthly Terrestrial Climatology. Journal of Climate. 1999;**12**:829-856

[21] Jones PD, Osborn JT, Briffa KR. Estimating sampling errors in large-scale temperature averages. Journal of Climate. 1997;**10**:2548-2568

[22] Jones PD. Hemispheric surface air temperature variations: A reanalysis and an update to 1993. Journal of Climate. 1994;**7**:1794-1802

[23] Kendall MG. Rank Correlation Methods. Charles Griffin & Company Limited. London, UK. 1948. 160 pp

[24] Peterson TC, Zhang X, Brunet-India M, Vázquez-Aguirre JL. Changes in North American extremes derived from daily weather data. Journal of Geophysical Research. 2008;**113**: D07113. DOI: 10.1029/2007JD009453

[25] Zhang X, Vincent LA, Hogg WD, Niitsoo A. Temperature and precipitation trends in Canada during the 20th century. Atmosphere-Ocean. 2000;**38**:395-429

[26] Burrows AT. The Chinook Winds. Journal of Geography. 1903;**2**(1903):124-136

[27] Giorgi F. Dependence of the surface climate interannual variability on spatial scale. Geophysical Research Letters. 2002;**29**(23):2101. DOI: 10.1029/2002GL016175

[28] Warrick RA. Drought in the Great Plains: A Case Study of Research on Climate and Society in the USA. Climatic Constraints and Human Activities. In: Ausubel J, Biswas AK, editors. IIASA Proceedings Series 10, 93-123. Pergamon Press, New York; 1980

[29] Jones PD, Moberg A. Hemispheric and large-scale surface air temperature variations: An extensive revision and update to 2001. Journal of Climate. 2003;**16**:206-223

[30] Lara MK, Snyder MA, Sloan LC. Irrigation cooling effect: Regional climate forcing by land-use change. Geophysical Research Letters. 2007;**34**:L03703. DOI: 10.1029/2006GL028679

[31] Kueppers LM, Snyder MA, Sloan LC, Cayan D, Jin J, Kanamaru H, Kanamitsu M, Miller NL, Tyree M, Du H, Weare B. Seasonal temperature responses to land-use change in the western United States. Global and Planetary Change. 2008;**60**:250-264

[32] Puma MJ, Cook BI. Effects of irrigation on global climate during the 20th century. Journal of Geophysical Research. 2010;**115**:D16120. DOI: 10.1029/2010JD014122

[33] Lobell DB, Bonfils C, Duffy PB. Climate change uncertainty for daily minimum and maximum temperatures: A model inter-comparison. Geophysical Research Letters. 2007; **34**:L05715

[34] Irmak S. Nebraska Water and Energy Flux Measurement, Modeling, and Research Network (NEBFLUX). Transactions of the ASABE. 2010;**53**(4):1097-1115

Chapter 10

Safe Drinking Water: Concepts, Benefits, Principles and Standards

Megersa Olumana Dinka

Additional information is available at the end of the chapter

http://dx.doi.org/10.5772/intechopen.71352

Abstract

Water is connected to every forms of life on earth. As a criteria, an adequate, reliable, clean, accessible, acceptable and safe drinking water supply has to be available for various users. The United Nation (UN) and other countries declared access to safe drinking water as a fundamental human right, and an essential step towards improving living standards. Access to water was one of the main goal of Millinium Development Goals (UN-MDGs) and it is also one of the main goal of the Sustainable Development Goals (SDGs). The UN-SDG goal 6 states that *"Water sustains life, but safe clean drinking water defines civilization"*. Despite these facts, there are inequalities in access to safe drinking water in the world. In some countries, sufficient freshwater is not available (*physical scarcity*); while in other countries, abundant freshwater is available, but it is expensive to use (*economic scarcity*). The other challenge is the increasing population of the world at an alarming rate, while the available freshwater resources almost remains constant. This chapter presents aspects of safe drinking water - background information, definition of water safety and access, benefits, principles and regulations, factors challenging the sustainable water supply and water quality standards and parameters.

Keywords: accessibility, inequalities, quality standards, safe water, water uses

1. Introduction

Water covers more than two-thirds of the earth's surface, but mostly salty and undrinkable. The available freshwater resource is only 2.7% of the available water on earth but only 1% of the available freshwater (in lakes, rivers and groundwater) is accessible. Most of the available freshwater resources are inaccessible because they are in the hidden part of the hydrologic cycles (deep aquifers) and in glaciers (frozen in the polar ice), which means safe drinkable water on earth has very small proportion (~3%) in the freshwater resources. Freshwater can also be obtained from the seawater by desalinization process. In some countries, sufficient freshwater is not available (*physical scarcity*). In some countries, abundant freshwater is available, but it is expensive to use (*economic scarcity*).

South Africa receives about 450 mm annual rainfall and is classified as a water-stressed country [1, 2]. The available freshwater resource can sustain 80 million people only. Some African countries (Ethiopia, Congo and Papua New Guinea) have excess freshwater resources, but they are having water shortage due to economic reasons. Ethiopia, the second populous countries in Africa, is the water tower of east Africa due to the availability of abundant water (nine major river basins). However, the country is among the few countries in the world affected by chronic water problem. The water scarcity in the world is further aggravated by the reduced water quantity (or an increased water demands) due to population growth and the declining of water quality by pollution.

As a criterion, an adequate, clean and safe drinking water supply has to be available for various users [3]. There is no universally accepted definition of "safe drinking water." Safe drinking water is defined as the water that does not represent any significant risk to health over a lifetime of consumption [4]. The safe drinking water must be delivered that is pure, wholesome, healthful and potable. Safe water is not necessarily pure, it has some impurities in it. It contains some traces of salts such as magnesium, calcium, carbonates, bicarbonates and others. The degree of purity and safety is a relative term and debatable. Clean/pure water has no minerals and it only contains H and O. According to the Monitoring organizations under the supervision of the Joint Monitoring Programme (JMP), "safe drinking water" is defined as water from an "improved water source," which includes household connections, public standpipes, boreholes, protected dug wells, protected springs and rainwater collections. According to the same organization, "access to safe drinking water" is defined as the availability of at least 20 l per person per day from an "improved" source within 1 km of the user's dwelling.

Safe drinking (potable) water is the water that can be delivered to the user and is safe for drinking, food preparation, personal hygiene and washing [3]. The water must meet the required (chemical, biological and physical) quality standards at the point of supply to the users [5]. Therefore, safe drinking water is a relative term, which depends on the standards and guidelines of a country; the standards set for the different quality parameters are different. The standard of WHO is not exactly the same as that of USA, Canada, European Commission, Russia, India, South Africa, Ethiopia, and so on. The term "safe" depends on the particular resistance ability of an individual. Water that is safe for drinking in some African countries might not be safe in European countries. Some African countries already developed resistance to some of the water-related diseases.

Safe drinking water is anonymously accepted as an international agenda and priority, which is evident from the MDGs and SDGs of the United Nations (UN) initiative and vision (MDGs 7 and SDGs 6). Despite the MDGs effort, still many people lack access to safe drinking water, even lack access to basic water. Globally, more than 1 billion people do not have access to safe drinking water. According to the Third World Academy of Sciences (TWAS) report [6], contaminated/dirty water is killing more people than cancer, AIDS, wars or accidents. Population of the world is increasing and the available freshwater resources almost remain constant.

http://dx.doi.org/10.5772/intechopen.71352

The number of people without access to safe drinking water is increasing. This is mostly related to the ever-increasing population growth in the developing countries and the inability (or unwillingness) of governments (local and national) to provide adequate water supply facilities in these countries [7].

2. Drinking water safety and access

2.1. Access to safe drinking water

Water is connected to every form of life on earth and is the basic human need, equally important as air. Water is connected to every aspect of human day-to-day activities directly or indirectly. At a basic level, everyone needs access to safe water in adequate quantities for drinking, cooking, personal hygiene and sanitation facilities that do not compromise health or dignity. Therefore, access to safe and dependable (clean and fresh) water is the fundamental/basic right of humans [8]. The UN and other countries declared that access to clean, safe drinking water is a basic human right, and an essential step toward improving living standards worldwide. Access to water was one of the main goals of UN-MDGs and it is also one of the main goals of the UN-SDGs. The South African constitution declares "*access to water and food for all*" as the main goal in the constitution following the 1998 National Water Act [9]. Despite these facts, still there are inequalities in access to safe drinking water in South Africa and in the world, the problem has more impacts on the poor, women and children. There are also inequalities within and among nations [6]. For instance, the population with access to safe drinking water in Congo was 77% for rural dwellers and 17% for rural dwellers by the year 2002 [6]. Inequalities in access to water and sanitation are morally unacceptable, but they are prohibited under international law [3].

Globally, it is estimated that 89% of people have access to water suitable for drinking [10]. According to UNDP [11] report, one out of six people do not have access to clean water, that is, about 1.1 billion people lack access to safe drinking water. In some countries, especially in Africa, almost half of the population do not have access to safe drinking water and hence, is afflicted with poor health [12]. The number of people without safe drinking water is more than the number reported by UNDP [11]. This is due to the fact that most of the water supply facilities initiated during the MDGs in developing countries are not functioning properly.

2.2. Benefits of safe drinking water

Water of satisfactory quality is the fundamental indicator of health and well-being of a society and hence, crucial for the development of a country. Contaminated water not only has the potential to pose immediate threat to human, but also can affect an individual productive rate [13]. According to the WHO [14] report, an estimated 1.1 billion people in the world drink unsafe water. Approximately 3.1% of the global annual death (1.7 million) and 3.7% of the annual burden (disability) (54.2 million) are caused by the use of unsafe water and lack of basic sanitation and hygiene.

Water provides a number of benefits and services for humans and the ecosystem. As reported by OECD [15], the benefit of water is not documented sufficiently, resulting in low political priority for water issues and in suboptimal levels of investment in water infrastructures. The same document also indicates that the benefit of water is mostly hidden in other technical documents. Most researchers have indicated that the benefit-cost ratio of access to water is more than 2, and in some cases, it can reach 7.0. In developing countries like Africa, the benefit-cost ratio of access to water is very high (more than 5:1 ratio) because it is related to every dimension of developmental activities (agriculture, energy, industry, etc.). In such areas, the return on investment in water services usually result in a substantial economic gains, estimated in the range of 5–28 USD per 1 USD [7]. In addition to the economic gains, water supply projects have technical, environmental and political gains. Water sector is interconnected with other development sectors (agriculture, energy, industry, etc.) and factors (social, economic, environmental, health, educational, legal and political) at local, national levels, regional and international levels [16]. In fact, access to safe water has a number of direct and indirect benefits related to health, education, poverty and environment. The UN World Water Development Report [7] indicated that there is a linkage or nexus between water and sustainable development, far beyond its social, economic and environmental dimensions. The report clearly indicated that access to safe water has a great role in addressing the developmental challenges, such as human health, food and energy security, urbanization and industrial growth, as well as climate changes. Especially, there is a strong nexus between water, food and energy [3].

The MDGs of the UN targeted to *"halve the population without access to safe drinking water and basic sanitation"* in the period from 1990 to 2015. According to the report by WHO and UNICEF [17] through their Joint Monitoring Programme (JMP) for water supply and sanitation, about 2.3 billion people have gained access to an improved drinking water. The report indicates an impressive gain has been made in the past two decades, but much has to be done. The success of MDGs is even doubtful since many of developing countries, especially the poor are still struggling to get access to safe drinking water. As stated in Section 2.1, the number of people without access to safe drinking water is more than the value reported by the UN.

Research has shown that the majority of people without access to safe water are from developing nations [18]. This shows that many people in the developing world, especially Africa, still depend on unsafe water sources for daily water need and affected by chronic water problems and water-borne diseases. Millions of people die due to water-related diseases like cholera, diarrhea, malaria, dengue fever, and so on. Globally, water-borne diseases kill more than 25,000 people per day and about 5000 children die per day due to water-related diseases (mainly diarrhea) [12], most of them can be easily prevented. Diarrhea and related diseases kill about 1.8 million children every year, most of them are in developing countries [19]. It is also estimated that about 1.8 billion people drink water contaminated with *Escherichia coli* (indicator of fecal contamination) [20]. In many parts of the world, especially developing countries, water-borne diseases represent the leading cause of death. Thus, access to safe water means a reduction of water-related diseases. It is an opportunity for improved health because it reduces the outbreak of health hazards.

In cognizant to the benefits of water, the newly introduced ambitious Sustainable Development Goal (SDG) by UN in 2014 [21] considers water as one of the main developmental pillars under SDG 6. In fact, water was also one of the main goals of the UN-MDGs. The UN-SDG 6 states that *"Water sustains life but safe, clean drinking water defines civilization."* The UN-SDG 6 recommended a dedicated SDG for water under five target areas such as (i) WASH, (ii) water resources, (iii) water governance, (iv) water quality and wastewater management and (v) water-related disasters. This indicates that the benefit-cost ratio of water is very high since it has social, economic, financial and environmental benefits. The benefit of water extends to other developmental activities/sectors such as health, education, agriculture and food production, energy, industry and other social and economic activities [7]. Therefore, achieving the UN's SDG 6 seems very hard, especially in the poorest countries like Africa where there are lots of problems and challenges. It requires dramatic improvement to the quality of life and longevity [7]. If we declare that "access to clean safe drinking water is a basic human right, then providing the necessary education, infrastructure and support to ensure the success of SDG 6 is the responsibility of us all." In developing countries, improving access to safe water requires the establishment of good governance [22].

3. Basic principles of safe drinking water supply

3.1. Definition of terms

There are basic standards, norms, criterion and indicators for safe drinking water. There are also policies, strategy and program under safe drinking water. These terms are well defined by Bos et al. [3]. Norm refers to the standard of development related to the large group of society. Criterion refers to the agreed norm or standard used for the decision. Indicator refers to the measured value of individual water quality parameters. Standard refers to the agreed target/threshold value established as an agreed target, which is set by an authority. There are various water quality standards and criteria in the world. Details of the water quality standards are provided under Section 5.3.

3.2. Water regulations and act

Water regulations are important for the provision of drinking water that is sufficient in quantity, safe, accessible, acceptable, affordable and reliable. Drinking water regulations include controlling of the water supply systems which are water source, water treatment, distribution, use, wastewater and gray water. Countries regulate drinking water differently depending on the quality of their water source. As stated earlier, different countries regulate drinking water differently depending on the quality of their water source.

In South Africa, water sources are monitored by the Department of Water and Sanitation (DWS). This was achieved by the implementation of the National Water Act (NWA) 36 of 1998 [9]. The purpose of the NWA is to ensure that the nation's water resources are protected, used, developed, conserved, managed and controlled. Local authorities are responsible for the supply of water to residents. This was achieved by the implementation of the Water Services Act (WSA) 108 of 1997. WSA are established to provide the following services [9]: (1) ensuring the rights of access to basic water supply and sanitation; (2) setting national standards, norms and tariffs; (3) water service development plans; (4) prepare the regulatory framework for water service institutions and intermediaries; (5) establish and disestablish committee for water boards and water services and their powers and duties; (6) monitoring water services and intervention and (7) providing financial assistance to water service institutions.

As a criterion, an adequate, clean and safe drinking water supply has to be available for various users [3]. Moreover, water has to be accessible for all, including children, elders and disabled ones. Water **availability** refers to both sufficient quantities and reliability of service provisions. **Adequacy** refers to both the quality and quantity of water. **Reliability** refers to continuity of the service provision for the current and future generation, which is covered under the principle of sustainability, system robustness and resilience. **Acceptability** refers to esthetic value of water – the acceptable appearance, taste and odor of water. It is highly subjective parameter and largely depends critically on the perceptions of the local ecology, culture, education and experience and hence, there is no set clear and objective global acceptability standards. **Accessibility** to water refers to the accessibility to a reliable supply of water on a continuous basis close to the point of demand: within everyone's reach: home, school, work, public places. It is related to the distance of water source from the point of demand (30 minutes walk or 0.2 km). That means the water has to be accessible for everyone, including children, elders and disabled ones. The detailed definition of the above water variables can be obtained from Bos et al. [3].

The role of a drinking water supplier is to provide adequate water for the community and prevent/mitigate risk of water contamination in different elements/points of water supply system such as source, treatment and distribution. They also should assure the delivery of a safe and esthetically pleasing drinking water to the consumer's point. In general, the prevention, mitigation and elimination of water contamination are the responsibilities of water providers and regulators. Water regulations are also important for the provision of drinking water that is sufficient in quantity, safe, accessible, acceptable, affordable and reliable. Countries regulate drinking water differently depending on the quality of their water source. According to the WHO [23] and US Environmental Protection Agency [24], there are guidelines and principles that need to be followed for water to be considered fit for use. The guidelines are as follows: physical, microbial, chemical and radiological. The water quality standards for different countries are summarized under Section 6.1.

4. Potential factors challenging water supply systems

The water supply system (WSS) is a system of hydrologic and hydraulic components, including all buildings and installations, used to meet water requirement of industrial and population centers. It consists of capturing raw water, drainage basin, water capturing and transmission pipes, water treatment plants, treated water transfer pipes, drinking water adduction pipes, pumping stations and pumping, water storage tanks and water distribution networks to the consumers [25–27]. A conventional water supply system is a combination of complex subsystems, consisting of the water supply catchment, water storage reservoir, water treatment plant and water distribution network [26]. Water supply and distribution systems typically comprise a combination of source works, treatment facilities, service reservoirs, pumping stations, pipes, valves and so on [25].

4.1. Sustainable water supply and challenges

In the ambitious vision 2050 of the SDG, sufficient and safe water has to be available for all to support human's basic needs and ecosystem integrity [7]. The sustainable development of the world largely depends on the sustainable development of water since other sectors are interrelated with water resources. It requires the progress of the three dimensions of the sustainable development (social, economic and environmental) [7]. Thus, the vision of SDGs (goal 6) for water requires management of the available water and related resources in an integrated, inclusive and participatory approach. Huge investment is highly needed for infrastructure, treatment plant systems and water recycling [29].

A WSS may face a number of challenges associated with many factors in provision of quality, efficient, reliable, resilient and sustainable water supply for the present and future generations. Rural areas are facing more financial and technical difficulties than urban areas. According to da Silva et al. [29], wealthier urban areas have more financial capacity and technical expertise than the poor rural communities to raise the capital needed for water infrastructure. Especially in rural areas with arid environment and great hydrologic variability, reliable and dependable WSS requires energy intensive infrastructure. A study made by Chung et al. [30] showed that robust optimization approach is a useful tool in reliable WSS design, under uncertainty, that prevents system failure at a certain level of risk.

Achieving the SDG requires huge **capital investment and good governance**, which is lacking in developing countries. Huge investment is highly needed for infrastructure, treatment plant systems and water recycling [28]. The sustainable development of water sector is affected by the sustainable development of the other sectors. Unsustainable developmental activities are greatly threatening the quantity and quality of renewable freshwater resources. Various driving forces are threatening the sustainability of WSS such as population increase at alarming rate, high rate of urbanization, significant land cover and climate change, the high demand for new energy supplies and poor governance. These driving factors are causing an increasingly frequent water shortage, floods and droughts, deleterious runoff, coastal hypoxia and depleted aquifers [28]. They have challenged the success of MDGs and will continue challenging the achievement of the newly set MDGs.

The other challenge of sustainable water supply is the **lack of appropriate policies and programs that consider rural diversity.** Small rural communities are the most vulnerable to water contamination. Furthermore, they struggle to secure the necessary funds for infrastructure necessary to improve water treatment and delivery systems, and thus fail to meet drinking water quality regulations. Community management is the tendency to provide water to rural areas worldwide. Despite the diversity of rural communities and their water supplies, policies tend to be uniform. A quantitative and qualitative study made in the Colombian Andes on four rural water supplies by considering aspects of infrastructure, training of human resources, revenue collection, water quality and post-construction support [31]. The study concluded that there is a need to design policies and programs that consider rural diversity to facilitate the sustainable water supply services. According to Kot et al. [32], policymakers have to align small communities with appropriate water quality goals by considering the contextual and cultural differences among rural communities.

In urban areas, the infrequent and insufficient application of **adaptive capacity indicators in urban sustainable water supply** systems has led to the challenge of dynamic and uncertain urban water supply systems. This condition is threatening the sustainability of urban water supply systems and raises concerns about the progress of urban water systems for variation and change [33]. As suggested by Spiller [33], future research should focus on developing methods and indicators that can define, evaluate and quantify adaptive capacity indicators under the three dimensions of sustainable development (*economic, environmental and technical*). Therefore, there is an urgent need to move toward the use of adaptive capacity indicators.

Moreover, there is an urgent need to move toward **sustainable and resilient smart water grids** in urban areas. Urban water supply systems are facing challenges of sustainability and resiliency, including water leaks, over-use, quality issues and response to drought and natural disasters [34]. Information and communications technology could help address these challenges through the development of smart water grids that network and automate monitoring and control devices [34]. While impressive progress has been made on technological elements (information and communication), the application of a smart water grid has received scant attention, especially in developing countries.

In fast-growing urban regions, **water demand and supply modeling** is extremely important. An **accurate prediction** of water demand plays a crucial role for water service providers in the planning, design and water utility asset management of drinking WSS. However, accurate prediction is always challenging due to the fact that predicting models require a simultaneous consideration of a number of factors affecting water demand and supply pattern. Some of the factors include climate changes, economic development, population growth, migration and consumer behavioral patterns [35].

4.2. Challenging factors for water supply systems

There are a number of factors challenging WSS. Some of the factors are aging infrastructure, water service provision thinking horizons, catchment (mountain)-specific issues, climate change, knowledge gaps with respect to present and future hydrology, accurate water demand prediction, land use/cover change, optimal operation of water supply systems, cost recovery, operating cost, water quality (water pollution), water scarcity, water leaks, low water pressure, over-use, response to drought and natural disasters, rapid urbanization, population growth, migration, demographic changes, economic development, consumer behavioral patterns, efficiency and reliability of a water supply system, self-sufficiency through use of alternative water sources, dynamic and uncertain urban water systems, complex dynamic human-environment coupled systems (non-holistic or siloed management), lack of adaptive capacity indicators to assess sustainability of water systems, scant attention of smart water grids (not supported by information and communications technology), lack of policies and programs that consider rural diversity and cultural differences and neglecting wastewater management are mentioned as challenges to water supply systems for provision of sustainable and reliable water services, which meet acceptable standards for present and future generations [14, 25, 26, 28–49].

According to Berg and Danilenko [38], WSS has faced a number of global challenges in the twenty-first century. The major challenges are population growth, uncertain climate changes, socio-environmental issues, limited water resources, economic crises and continuous aging process. There are a number of problems associated with the continuous aging process, including low pressure, water loss and water quality deterioration [36]. The major challenges in the provision of safe water and sanitation on a global basis are [37]: (1) water contamination within distribution systems; (2) increasing water scarcity and shortages; (3) implementing innovative and low-cost sanitation systems; (4) providing sustainable water supply systems and sanitation for megacities; (5) reducing the disparities in access to water and sanitation and (6) developing financially feasible water and sanitation services.

Increasing urban water self-sufficiency: The main drivers for increased self-sufficiency were identified to be direct and indirect lack of water, constrained infrastructure, high-quality water demands and commercial and institutional pressures. Public water service providers should plan to achieve a high level of reliable, stable and dependable water supply, which can be achieved by combining alternative water supply systems with the conventional ones. A case study made by Rygaard et al. [39] demonstrated an increase in water self-sufficiency ratios to more than 80% when the conventional water supply was supplemented by water recycling, seawater desalination and rainwater harvesting. However, the study indicated that care should be made during the introduction of alternative freshwater sources since it may raise several challenges such as very high-energy requirements (>*tenfold*) by the alternative techniques, appearance of trace contaminants in recycled wastewaters and the possible resistance from consumers due to the changes made to the drinking water system. The study concluded that despite the challenges, urban water self-sufficiency concepts in combination with conventional water resources are already helping to reach the goal of urban WSS.

Infrastructure development: Water services are in crisis or approaching crisis conditions due to the neglect of infrastructure, particularly underground water mains and sewers, largely because of political unwillingness to allow charges to be set high enough to achieve sustainable cost recovery. This is true in both developed and developing countries [43]. In developed countries, the solutions are relatively affordable; what is needed is the political commitment to take action. In developing countries, the situation is more serious due to a combination of neglect and rapidly growing urban populations. Without doubt, infrastructure is essential for sustainable water development. But infrastructure alone will not contribute to the improvement of the quality of life unless it is part of an overall framework: development, economic growth, social equity and environmental protection. As mentioned by the Nobel laureate Amartya Sen [45], "the absence of infrastructure has a pervasive influence on poverty, but at the same time is not a free-standing factor in lifting people from it." Thus, the focus should be the use of physical infrastructure as a driver for sustainable development. But infrastructure development takes more time beyond the life of most governments. The thinking of water service providers has to be based on long-term horizons. In order to improve the accountability and social welfare of relatively low-income households, there is a need for more comprehensive frameworks (institutional, legal, regulatory, policy and management) than the existing ones at present [45]. Venkatachalam [47] suggested that improving the existing public water supply to a satisfactory level will improve the household's willingness to pay because the willing households could reap significant benefits from the improved supply. This would help the government agencies to come out with an improved water tariff policy that will cover cost of investment and maintenance.

Urban water pricing (*cost recovery, affordability and water conservation*): Policymakers increasingly consider pricing as an important tool for cost recovery, affordability and water conservation to address water scarcity issues. However, implementing **tariff reforms** is often difficult in practice due to political factors and the absence of governance structures that can result in quality service provision. Additionally, institutional replication of successful water pricing policies has been difficult due to incomplete information and the contextual uniqueness of local institutions, politics and social relations. Water service provision thinking has to be based on long-term horizons. Infrastructure development takes time beyond the life of most governments. In those countries without such political continuity, there is a need for all political factions to agree on goals, policies and plans. It is unlikely that water can ever be separated from politics, but city political consensus must be attempted [53].

Climate change: Climate change is affecting the frequency of extreme weather events and hence increasing the uncertainty about water availability and reliability [50]. A properly planned, developed and managed infrastructure and related institutional capacities are required in order to buffer seasonal climatic variations and address water demand issues. More emphasis should be given to mountain-specific issues. Major priority areas include water governance for transboundary basins, cross-border information systems, establishing a knowledge base for mountain regions and sharing benefit between mountain and downstream communities [42].

Knowledge gaps: With respect to present and future, hydrology poses a serious constraint for infrastructure development. Changing hydrology will pose special challenges to the design, planning and management of infrastructure [42]. Land use influences raw surface water quality and treatment costs for drinking water supply [51]. Anthropogenic disturbances to the environment can compromise valuable ecosystem services, including the provision of potable water. These disturbances decrease water quality, potentially increasing treatment costs for producing drinking water.

Efficiency and reliability of a water supply system: Water inflow is among primary determinants of the successful functioning of the entire water supply system since it influences water storage. Developing an approach to assess the resilience of WSS under limited rainfall provides useful insights into effective system management [26]. For instance, understanding WSS resilience can support the identification of the minimum/threshold rainfall value by which WSS can maintain its operation without failure. It can also help to understand and identify the sensitivity of the WSS to a changing rainfall amount and distribution pattern.

In this regard, the water service providers are well aware of the stability of WSS and know when the system experience a pressure or disruptive influences.

Challenges for water supply and Governance: Cities struggling to keep pace with population and demographic changes are not unique. According to a study conducted in Dublin [41], collectively there are combinations of factors that create an inordinately challenging situation for those attempting to plan for the city's current and future water resources needs. Their main challenges related to topography, old infrastructure (the nineteenth century), population growth and development needs, water charges, climate change and water supply history.

5. Drinking water quality

5.1. Definition and concepts

Water is most fundamental in shaping the land and regulating the climate. It is one of the most important resources that profoundly influence life. Water quality is the most fundamental controlling factor when it comes to health and the state of diseases in both humans and animals. According to WHO report [23], about 80% of all the human diseases in human beings are caused by water.

Depending on the purpose of water quality analysis, water quality can be defined based on a set of biological, physical and chemical variable, which are closely linked to the water's intended use. As a principle, drinking water is supposed to be free from harmful pathogens and toxic chemicals [3]. Contamination of freshwater (especially groundwater) sources is one of the main challenges currently faced by the South Africans, more especially in communities who depend almost exclusively on groundwater [52]. Groundwater is used for domestic, industrial and agricultural water supply in all four corners of the world. Therefore, the presence of contaminants in natural freshwater continues to be one of the most important environmental issues in many areas of the world, more especially in developing countries [53]. Once the groundwater is contaminated, its quality cannot be restored back easily, the best way is to protect it.

The concept and theory of water quality is very broad since it is influenced by many factors. Water quality is based on the intended uses of water for different purposes, that is, different water uses require different criteria to be satisfied. In water quality analysis, all of the accepted and unaccepted values must be clearly defined for each quality variable. If the quality variables meet the pre-established standards for a given use is considered safe for that use. When water fails to meet these standards, it must be treated if possible before use.

5.2. Description of water quality parameters

5.2.1. Physical parameters

Physical quality parameters are related to total solids content, which is composed of floating matter, settleable matter, colloidal matter and matter in solution. The following physical parameters are determined in water [12]:

- **Color**: caused by dissolved organic materials from decaying vegetation or landfill leachate.
- **Taste and odor**: can be caused by foreign compounds such as organic compounds, inorganic salts or dissolved gases.
- **Temperatures**: the most desirable drinking water is consistently cool and does not have temperature fluctuation of more than a few degrees. Groundwater generally meets these criteria.
- **Turbidity**: refers to the presence of suspended solid materials in water such as clay, silt, organic material, plankton, and so on.

5.2.2. Chemical parameters

The chemical constituents have more health concerns for drinking water than for the physical constituents. The objectionability of most of the physical parameters are based on esthetic value than health effects. But the main objectionability of some of the chemical constituents is based on esthetic as well as concerns for adverse health effects. Some of the chemical constituents have an ability to cause health problems after prolonged period of time [54]. That means the chemical constituents have a cumulative effect on humans. The chemical quality parameters of water include alkalinity, biological oxygen demand (BOD), chemical oxygen demand (COD), dissolved gases, nitrogen compounds, pH, phosphorus and solids (organic). Sometimes, chemical characteristics are evidenced by their observed reactions such as in laundering, redox reactions, and so on [12, 54].

Below is a list of some of the chemical compounds and elements found in water:

- **Arsenic**: occurs naturally in some geologic formation. It is mostly used in agricultural chemicals in South Africa. In drinking water, it has been linked to lung and urinary bladder cancer.
- **Chloride**: most waters contain some chloride. The amount found can be caused by the leaching of industrial or domestic waters. Chloride should not exceed 100 mg/L in domestic water to be palatable.
- **Fluoride**: is a natural contaminant of water. It is one of those chemicals given high priority by WHO [14] for their health effects on humans. High F in drinking water usually causes dental and skeletal fluorosis. Excessive F (>2 mg/L) causes a dental disease known as fluorosis (mottling of teeth), while regular consumption in excess may give rise to bone and skeletal fluorosis [12]. On the other hand, F < 2 mg/L causes dental cavities in children.
- **Zinc**: is found in some natural waters, particularly in areas where zinc ore deposit have been mined. Though it is not considered detrimental to health, but it will impart a bad taste to drinking water.
- **Iron**: small amounts of iron frequently are present in water because of the large amount of iron in the geologic materials. This will cause reddish color to water.

- **Manganese**: naturally occurring manganese is often present in significant amounts in groundwater. Anthropogenic sources include discarded batteries, steel alloy production and agricultural products.
- **Toxic substances**: generally classified as inorganic substances, organic substances and heavy metals. The toxic inorganic substances include nitrates (NO_3), cyanides (CN_) and heavy metals. These substances are of major health concern in drinking water. High NO_3 content can cause *Methemoglobinemia* in infants ("infant cyanosis" or "blue baby syndrome"); while CN can cause oxygen deprivation [12]. There are more than 120 toxic organic substances [24], generally exist in the form of pesticides, insecticides and solvents. These compounds produce health effects (acute or chronic). The toxic heavy metals are arsenic (As), barium (Ba), cadmium (Cd), chromium (Cr), lead (Pb), mercury (Hg), selenium (Se) and silver (Ag) [12]. Like the organic substances, some of these substances are acute poisons (As and Cr) and others produce chronic diseases (Pb, Cd and Hg).

5.2.3. *Biological parameters*

Biological parameters are the basic quality parameters for the control of diseases caused by pathogenic organisms, which have human origin. Pathogenic organisms found in surface water include bacteria, fungi, algae, protozoa, plants and animals and viruses. Some of these disease-causing organisms (bacteria, fungi, algae, protozoa and viruses) are not identifiable and can only be observed microscopically. Microbiological agents are very important in their relation to public health and may also be significant in the modification of physical and chemical characteristics of water [12]. Water for drinking and cooking purposes must be free from pathogens. The greatest microbial risks are associated with consumption of water that is contaminated with human or animal feces. Feces can carry pathogenic bacteria, protozoa, helminthes and virus. Pathogens originating from feces are the principle concerns in setting health-based targets for microbial safety. Water-borne diseases are particularly to be avoided because of the capacity of result in the simultaneous infection of large number of people. While water can be a very significant source of infectious organisms, many of the diseases that may be waterborne may also be transmitted by other routes, including person-to-person contact, droplets and aerosols and food intake [54].

The techniques for comprehensive bacteriological test are complex and time consuming. Different tests have been developed to detect the relative degree of bacterial contaminations in terms of an easily defined quantity. There are two mostly used test methods widely used to estimate the number of microorganism of coliform groups (*Escherichia coli and Aerobacter aerogenes*). These include: total coliforms or *E. coli*, but the second one is found to be a better indicator of biological contamination compared to the first one [12].

5.3. Water quality standards

As presented in Section 3.1, standard is defined as a basis for judging the quality. A standard for drinking water quality is thus the reference that will ensure that the delivered water will not pose any threat or harm to human health. The water quality standard is the framework against which a water sample can be considered satisfactory or safe for use [54]. There are a number of standard guidelines for drinking purposes such as World Health Organization [54], Commission for European Union [55], U.S. Environmental Protection Agent [24], Environmental Canada [56], Russian Standard [57], Indian Standard [58, 59], South African National Standard [60] and Ethiopian Standards [61]. Most developing and other developed countries use the WHO standards for drinking water [54]. **Table 1** summarizes water quality guidelines of different countries.

5.4. Water quality index

It is difficult to quantify the overall suitability of water for drinking based on the various guidelines presented in **Table 1**. The interpretation of the various water quality parameters separately is usually a difficult task for general public as well as decision and policy makers. Therefore, the calculation of a general water quality index (WQI) is extremely important in order to communicate the quality of water in a better and understandable ways. There are different approaches of calculating WQI. In this section, a brief description has been provided for the weighted Arithmetic Water Quality Index Method proposed by Tiwari and Mishra [62] and adopted by others [63–67]. The quality rating (q_i), the sub-index (SI) [65] and the relative weights (Wi) are calculated using Eqs. (1)–(3).

$$q_i = 100\left(\frac{V_i - V_{o,}}{S_i - V_i}\right) \quad (1)$$

$$SI_i = (W_i)(q_i) \quad (2)$$

$$W_i = \frac{w_i}{\sum w_i} \quad (3)$$

where V_i and S_i are the analytical and the standard value for the ith parameter, respectively, V_o is the ideal value of the ith parameter in pure water ($V_o = 0$, except pH =7.0). The standard value is usually considered as the maximum permissible level set by WHO [10, 14, 54] or as per the standards for different countries presented in **Table 1**. W_i is the relative weights for various water quality parameters, assumed to be inversely proportional to the recommended standards for the corresponding parameters. w_i is the unit weight of each parameter according to its relative importance in the overall quality of water for drinking purposes. The w_i values are provided by Tiwari and Mishra [62], which depend on the number of parameters considered in the calculation of WQI. Note that the $\sum W_i$ should be equal to 1.

Finally, the overall WQI (Eq. (4)) is calculated for each of the water sources by aggregating the quality rating (q_i) linearly and taking their weighted mean.

$$WQI = \sum_{i=1}^{n} SI_i \quad (4)$$

WQI classes are as follows: 0–25 (excellent, grade A), 26–50 (good, grade B), 51–75 (poor, grade C), 76–100 (very poor, grade D), >100 (unfit for drinking, Grade E).

Parameters	Standard concentrations								
	WHO[a]		USA (USEPA[b])	Europe (CEU[c])	Russia[d]	Canada (EC[e])	India[f]	South Africa[g] (SANS)	Ethiopia[h] (ESA)
	HDL	MPL							
PH	7.0–8.5	6.5–8.5	6.5–8.5	6.5–8.5	6.0–9.0	6.5–8.5	6.5–9.2	6.5–9.0	6.5–8.5
EC	300	1400	–	–	2000	–	–		–
Na^+	–	200	–	–	200	20	150	100	200
Ca^{2+}	75	100	–	–	–	–	100	32	75
Mg^{2+}	30	50	–	–	–	–	100	30	50
K^+	12	200	–	–	–	–	–	50	1.5
Cl^-	200	600	250	250	350	250	1000	200	250
CO_3^-	–	45	–	–	–	–	–	–	–
HCO_3^-	–	500	–	–	–	–	400	–	–
TH	200	500	–	–	–	300	600	–	300
F	1.0	1.5	2.0	1.5	1.5	1.5	–	1.0	1.5
B	–	0.3	–	1.0	0.3	5.0	5	–	0.3
SO_4^-	200	250	250	250	500	250	400	200	250
TDS	500	600	500	–	1000	500	1500	450	1000

P – probability (%); HDL – highest desirable limit; MPL – maximum permissible limit; USEPA – United States Environmental Protection Agency; CEU – Commission of European Union; EC – Environmental Canada.

Sources: [a]WHO [54], [b]USEPA [24], [c]CEU [55], [d]UNESCO/WHO/UNEP [56], [e]Health Canada [57], [f]ISI [58] and BIS [59], [g]SANS [60], [h]ESA [61]. Note that the values indicated for the different standards other than WHO are the maximum permissible limits.

Table 1. Comparison of the different drinking water standards.

6. Conclusion

As water is a basic need for human life, access to clean, safe drinking water is a basic human right. As a criterion, an adequate, reliable, clean, acceptable and safe drinking water supply has to be available for various users. Moreover, everyone needs access to safe water in adequate quantities for drinking, cooking and personal hygiene and sanitation facilities that do not compromise health or dignity. Access to water is one of the most important catalysts given high priority by the UN for sustainable development. Despite these facts, there are inequalities in access to safe drinking water in the world. There are a number of factors challenging the sustainable WSS. Some of the factors are related to infrastructures (aging), clean water issues (quality, scarcity), natural factors (climate change, flood and drought), human factors (population growth, migration, demographic change, economic development, willingness to pay for water supply services, overuse), water management and delivery problems (pressure, leakages, lack of smart water meters, cost recovery, operation costs, etc.).

MDG fails to achieve its goal for access to safe water and sanitation. The chance for the success of the newly set SDG is also not different from that of MDGs, especially in some African countries. Some of the African leaders are reporting a false number of people with access to safe drinking water and sanitation to get a donation from the UN and using the donated money to buy weapons and use it to suppress the right of the people. In developing countries, improving access to safe water requires provision of good quality education and the establishment of good governance. Priorities should be given to the development of a democratic government and community empowerment.

Author details

Megersa Olumana Dinka

Address all correspondence to: magarsol@yahoo.com

Department of Civil Engineering Sciences, Faculty of Engineering and the Built Environment, University of Johannesburg, South Africa

References

[1] Department of Water Affairs (DWA). Integrated Water Resource Planning for South Africa: A Situation Analysis. Pretoria: Department of Water Affairs; 2011

[2] Binns T, Illgner P, Nel E. Water shortage, deforestation and development: South Africa's 'Working for Water' programme. Land Degradation and Development, 2001;**12**:341-355

[3] Bos R, Alves D, Latorre C, Macleod N, Payen G, Roaf V, Rouse M. Manual on the Human Rights to Safe Drinking Water and Sanitation for Practitioners. London, UK: IWA Publishing; 2016

[4] Fogden J, Wood G. Access to Safe Drinking Water and Its Impact on Global Economic Growth. Bothell, WA, USA: A Study for HaloSource, Inc; 2009

[5] de Zuane J. Handbook of Drinking Water Quality. NY, USA: John Wiley & Sons; 1997

[6] Third World Academy of Sciences (TWAS). Safe Drinking Water: The Need, the Problem, Solutions and an Action Plan. Trieste, Italy: TWAS; 2002. p. 23

[7] United Nations Scientific and Cultural Organization (UNESCO). Water for a Sustainable World. The United Nations World Water Development Report. Paris, France: UNESCO; 2015

[8] Samra SCJ, Fawzi SCM. The right to water in rural Punjab: Assessing equitable access to water through the Punjab rural water supply and sanitation project. Health and Human Rights. 2011;**13**(2):36-49

[9] Department of Water and Sanitation (DWS). National Water Act No. 36. Republic of South Africa: DWS; 1998

[10] WHO 2017. Drinking Water Factsheet. Updated July 2017. Available from: http://www.who.int/mediacentre/factsheets/fs391/en/

[11] UNDP 2015. Sustainable Development Goal. 17 Goals to Transform Our World. Goal 6: Ensure Access to Water and Sanitation for All. UNDP Report. Available from: http://www.un.org/sustainabledevelopment/water-and-sanitation/

[12] Davis ML. Water and Wastewater Engineering: Design Principles and Practice. New York: McGraw-Hill Education; 2013

[13] Mpenyana-Monyatsi L, Momba MNB. Assessment of groundwater quality in the rural areas of the North West Province, South Africa. Scientific Research and Essays. 2012;**8**(7): 903-914

[14] WHO. UN-Water Global Annual Assessment of Sanitation and Drinking Water Report: The Changes of Extending Sustaining Services, UN Water Report 2012. Switzerland: WHO; 2012

[15] OECD 2011. Benefits of Investing in Water and Sanitation: An OECD Perspective. OECD; p. 148

[16] Biswas KA. Integrated water resources management: A reassessment a water forum contribution. International Water Resources Association. 2004,**29**(2).248-256

[17] World Health Organization/United Nations Children's Fund (WHO/UNICEF). Drinking Water: Equity, Safety and Sustainability. Geneva/New York: WHO/UNICEF; 2011

[18] United Nations World Water Assessment Programme (WWAP). The United Nations World Water Development Report: Water and Jobs. Paris: UNESCO; 2014

[19] Johnston RB, Berg M, Johnson CA, Tilley E, Hering JG. Water and anitation in developing countries: Geochemical aspects of quality and treatment. Elements. 2011;**7**:163-168

[20] Bain R, Cronk R, Hossain R, Bonjour S, Onda K, Wright J, Yang H, Slaymaker T, Hunter P, Prüss-Ustün A, Bartram J. Global assessment of exposure to faecal contamination through drinking water based on a systematic review. Tropical Medicine and International Health. 2014;**19**(8):917-927

[21] UN-Water 2014. A Post-2015 Global Goal for Water: Synthesis of Key Findings and Recommendations from UN-Water. Available from: http://bit.ly/Prg2lt

[22] Water Governance Facility (WGF). Human Rights-Based Approaches and Managing Water Resources: Exploring the Potential for Enhancing Development Outcomes. WGF Report No. 1. Stockholm: Stockholm International Water Institute (SIWI); 2012

[23] WHO/UNICEF. Progress on Drinking Water and Sanitation: 2013 Update. New York: Joint Monitoring Programme for Water Supply and Sanitation; 2013

[24] US EPA. Implementation and Enforcement of the Combined Sewer Overflow Policy. Report to Congress, EPA 833-R-01-003. Washington, DC: Environmental Protection Agency, Office of Water; 2001

[25] Rao ZF, Wicks J, West S. Optimising water supply and distribution operations. Proceedings of the Institution of Civil Engineers-Water Management. June 2007;**160**(2):95-101 Thomas Telford Ltd

[26] Amarasinghe P, Liu A, Egodawatta P, Barnes P, McGree J, Goonetilleke A. Quantitative assessment of resilience of a water supply system under rainfall reduction due to climate change. Journal of Hydrology. 2016;**540**:1043-1052

[27] Leitner I, Matuz B, Dippong T. Price analysis of the components of water supply systems. Scientific Bulletin Series D: Mining, Mineral Processing, Non-Ferrous Metallurgy, Geology and Environmental Engineering. 2016;**30**(2):45

[28] Sedlak DL, Schnoor JL. The challenge of water sustainability. Environmental Science Technology, 2013;**47**(11):5517

[29] da Silva EFO, Heikkila T, de Souza Filho FDA, Costa da Silva D. Developing sustainable and replicable water supply systems in rural communities in Brazil. International Journal of Water Resources Development. 2013;**29**(4):622-635

[30] Chung G, Lansey K, Bayraksan G. Reliable water supply system design under uncertainty. Environmental Modelling & Software. 2009;**24**(4):449-462

[31] Domínguez-Rivera I, Oviedo-Ocaña ER, Restrepo-Tarquino I. Service provision in rural water supplies: Analysis of four community-based systems in Colombia. Cuadernos de Desarrollo Rural. 2016;**13**(77):117-140

[32] Kot M, Gagnon GA, Castleden H. Water compliance challenges: How do Canadian small water systems respond? Water Policy. 2015;**17**(2):349-369

[33] Spiller M. Adaptive capacity indicators to assess sustainability of urban water systems – Current application. Science of the Total Environment. 2016;**569**:751-761

[34] Mutchek M, Williams E. Moving towards sustainable and resilient smart water grids. Challenges. 2014;**5**(1):123-137

[35] Qi C, Chang NB. System dynamics modeling for municipal water demand estimation in an urban region under uncertain economic impacts. Journal of Environmental Management. 2011;**92**(6):1628-1641

[36] Alegre H, Baptista JM, Cabrera E Jr, Cubillo F, Duarte P, Hirner W, Merkel W, Parena R. Performance Indicators for Water Supply Services. IWA, London; 2006

[37] Moe CL, Rheingans RD. Global challenges in water, sanitation and health. Journal of Water and Health. 2006;**4**(S1):41-57

[38] Berg C, Danilenko A. The IBNET Water Supply and Sanitation Performance Blue Book, The International Benchmarking Network for Water and Sanitation Utilities Data book, Water and Sanitation Program. Washington DC: The World Bank; 2011. p. 58849

[39] Rygaard M, Binning PJ, Albrechtsen HJ. Increasing urban water self-sufficiency: New era, new challenges. Journal of Environmental Management. 2011;**92**(1):185-194

[40] Haider H, Sadiq R, Tesfamariam S. Performance indicators for small-and medium-sized water supply systems: A review. Environmental Reviews. 2013;**22**(1):1-40

[41] Kelly-Quinn M, Blacklocke S, Bruen M, Earle R, O'Neill E, O'Sullivan J, Purcell P. Dublin Ireland: A city addressing challenging water supply, management, and governance issues. Ecology and Society. 2014;**19**(4):1-13

[42] Molden DJ, Vaidya RA, Shrestha AB, Rasul G, Shrestha MS. Water infrastructure for the Hindu Kush Himalayas. International Journal of Water Resources Development. 2014; **30**(1):60-77

[43] Rouse M. Policy brief: The urban water challenge. International Journal of Water Resources Development. 2013;**29**(3):300-309

[44] Rouse M. The worldwide urban water and wastewater infrastructure challenge. International Journal of Water Resources Development. 2014;**30**(1):20-27

[45] Tortajada C. Water infrastructure as an essential element for human development. International Journal of Water Resources Development. 2014;**30**(1):8-19

[46] Ma XC, Xue X, González-Mejía A, Garland J, Cashdollar J. Sustainable water systems for the city of tomorrow – A conceptual framework. Sustainability. 2015;7(9):12071-12105

[47] Venkatachalam L. Informal water markets and willingness to pay for water: A case study of the urban poor in Chennai City, India. International Journal of Water Resources Development. 2015;**31**(1):134-145

[48] United Nations World Water Assessment Programme (WWAP). The UN World Water Development Report: Wastewater Untapped Resource. Paris: UNESCO; 2017

[49] Donoso G. Urban water pricing in Chile: Cost recovery, affordability, and water conservation. Wiley Interdisciplinary Reviews: Water. 2017;**4**(2), e1194-n/a. DOI: 10.1002/wat2.1212

[50] United Nations World Water Assessment Programme (WWAP). The United Nations World Water Development Report: Water and Jobs. Paris: UNESCO; 2016

[51] Cunha DGF, Sabogal-Paz LP, Dodds WK. Land use influence on raw surface water quality and treatment costs for drinking supply in São Paulo state (Brazil). Ecological Engineering. 2016;**94**:516-524

[52] Engelbrecht JFP, Tredoux G. 2000. Bacteria in "unpolluted" groundwater. WISA 2000 Biennial Conference, 28 May - 1 June 2000, Cape Water Programme, CSIR, Stellenbosch, Sun City, South Africa

[53] Amaliya NK, Kumer SK. Evaluation of surface water quality of Kanyakumari district through water quality index assessment. International Journal of Plant, Animal and Environmental Sciences. 2013;**4**(1):73-77

[54] WHO. Guidelines for Drinking-Water Quality. 4th ed. Geneva, Switzerland: World Health Organization; 2011

[55] Council of European Union (CEU). Drinking Water Directive. Council Directive of 15 July 1980 related to the quality of water intended for human consumption, 80/778/EEC. CEU, Brussels; 1998. p. 32

[56] Health Canada. Guidelines for Canadian Drinking Water Quality. Canada: Federal-Provincial-Territorial Committee on Drinking Water; 2008

[57] UNESCO/WHO/UNEP. Water Quality Assessment – A Guide to Use of Biota, Sediments, and Water in Environmental Monitoring. 2nd ed. Cambridge, Great Britain: University Press E&FN Spon; 1996. p. 609

[58] Indian Standard Institution (ISI), Drinking Water Standard Substances or Characteristic Affecting the Acceptability of Water for Domestic Use. (IS: 105001983). Indian Standard Institute, India, p. 1-22

[59] BIS. Bureau of Indian Standards, IS: 10500. New Delhi, India: Manak Bhawan; 1991

[60] SANS 241 2006. South African National Standard – Drinking Water (6th ed.). Standards South Africa, SANS 241, WRC, Pretoria, ISBN: 0-626-17752-9

[61] Ethiopian Standard Agency (ESA). Drinking Water – Specifications. Ethiopian Standard Agency, ICS, Addis Ababa; 2013. p. 13.060.20

[62] Tiwari TN, Mishra M. A preliminary assignment of water quality index of major Indian rivers. Indian Journal of Environmental Protection. 1985;**5**(4):276-279

[63] Gupta DM, Purohit KM, Jayita D. Assessment of drinking water quality of river Brahmani. Journal of Environmental and Pollution. 2001;**8**:285-291

[64] Asadi SS, Vappala P, Reddy AM. Remotely sensing and GIS techniques for evaluation of groundwater quality in municipal Corporation of Hydrabad (zone-V), India. International Journal of Environmental Research and Public Health. 2007;**4**(1):45-52

[65] Ramakrishnaiah CR, Sadashivaiah C, Ranganna G. Assessment of water quality index for the groundwater in Tumkur Taluk, Karnataka State, India. E-Journal of Chemistry. 2009;**6**(2):523-530

[66] Dinka MO. Analyzing the Extents of Basaka Lake Expansion and Soil and Water Quality Status of Matahara Irrigation Scheme, Awash Basin (Ethiopia). [dissertation]. Vienna, Austria: University of Natural Resources and Applied Life Sciences; 2010

[67] Jagadeeswari P, Ramesh K. Water quality index for assessment of water quality in South Chennai coastal aquifer, Tamil Nadu, India. International Journal of ChemTech Research. 2012;**4**(4):1582-1588